THE  TIMES

Train Tracks

Book
5

THE TIMES

Train
Tracks

Book
5

200 challenging visual Train Track puzzles

Published in 2022 by Times Books

HarperCollins Publishers   HarperCollins Publishers
Westerhill Road   1st Floor, Watermarque Building
Bishopbriggs   Ringsend Road, Dublin 4, Ireland
Glasgow
G64 2QT
www.harpercollins.co.uk

10 9 8 7 6 5 4 3 2 1

© HarperCollins Publishers 2022

All individual puzzles copyright Puzzler Media - www.puzzler.com

The Times® is a registered trademark of Times Newspapers Limited

ISBN 978-0-00-853585-8

Layout by Puzzler Media

Printed and bound in the UK using 100% Renewable Electricity at CPI Group
(UK) Ltd

The contents of this publication are believed correct at the time of printing.
Nevertheless the publisher can accept no responsibility for errors or omissions,
changes in the detail given or for any expense or loss thereby caused.

A catalogue record for this book is available from the British Library.

If you would like to comment on any aspect of this book, please contact us at
the above address or online.
E-mail: puzzles@harpercollins.co.uk

This book is produced from independently certified FSC™ paper
to ensure responsible forest management.

For more information visit: www.harpercollins.co.uk/green

Contents

Solutions

Introduction

Introduction

A train is departing from town A and travelling to town B, but the track has not been completed. Can you find out where the rails must go?

The numbers at the top of each column and the end of each row tell you how many cells contain sections of track in that column or row. As in the 4 diagrams below, the train track can move straight through a cell horizontally or vertically, or curve to the left or right only. The track cannot cross itself.

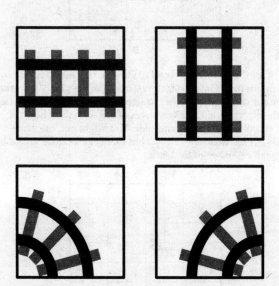

Work has already begun and some pieces of rail have already been placed in the grid to help you get started.

This is all you need to know to complete the track, so go to puzzle 1 to start building your train tracks or continue reading, here, for some simple solving tips.

Tip 1

Throughout your solving, it helps to mark cells where pieces of rail cannot go with an X.

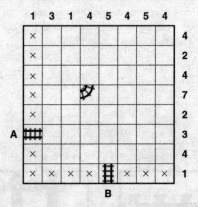

Tip 2

Except for the start (A) and end (B) towns, the train track is one continuous path that travels through cells in the grid. As the track enters any cell it can go either straight on or turn left or turn right. Use the direction of the given pieces of rail to draw in how the track enters its next cells either side. You might not know how the track will exit these new cells yet but more clues can be gained by this small extension of the track.

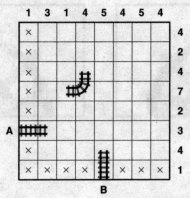

Tip 3

It is also important to consider the functionality of the pieces of rail you use. For example, when a turning piece is used it automatically uses up a minimum of two cells in that row or column – see the curved piece in the fourth row and column in the example from Tip 2.

So, by deduction, a straight piece of rail is the only piece that can be used when there is only one piece of rail allowed in a row or column. For example see column three, here.

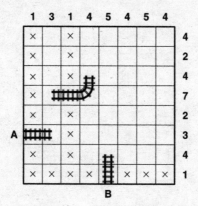

These three tips will set you safely on your way to expert rail construction. Don't forget, whatever the problem, the track needs to keep on going and get the train to its destination.
Enjoy the journey.

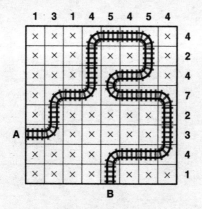

Puzzles

Easy Train Tracks

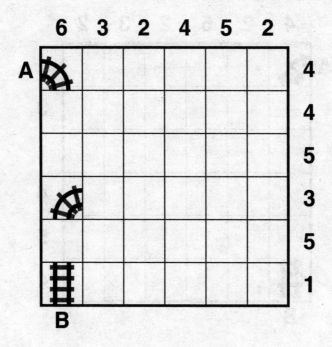

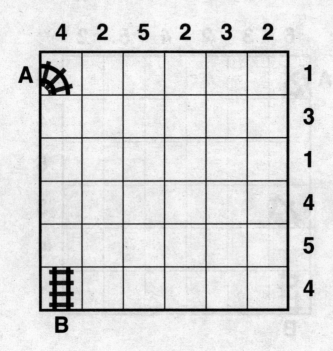

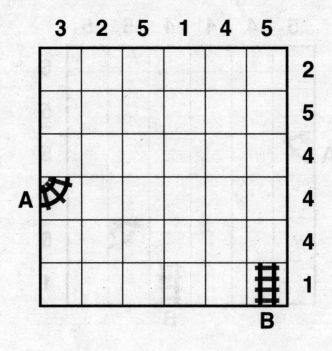

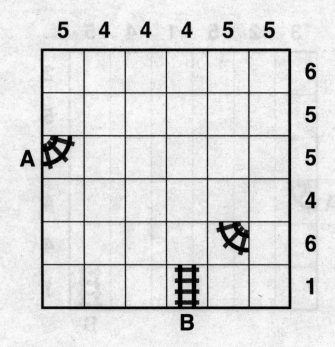

5

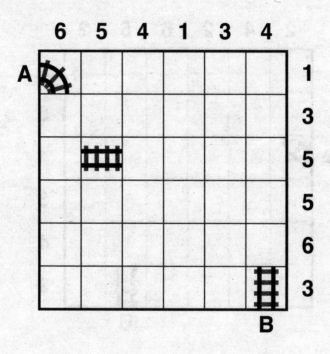

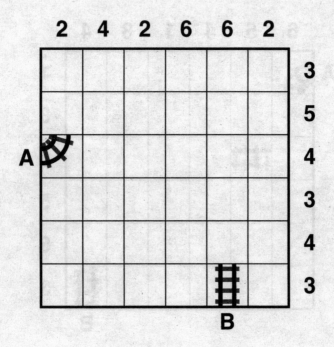

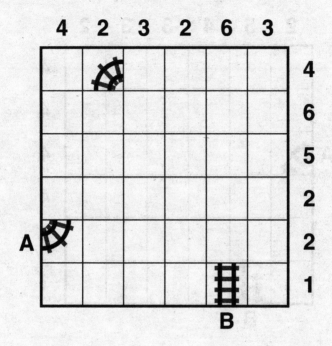

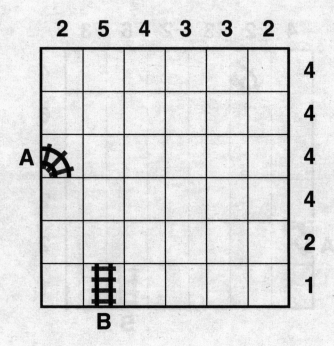

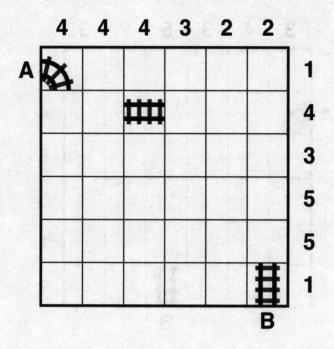

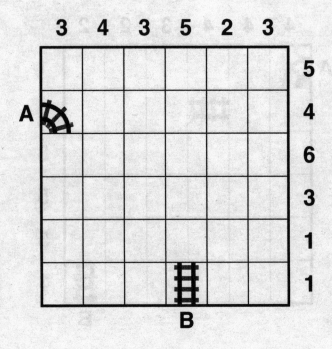

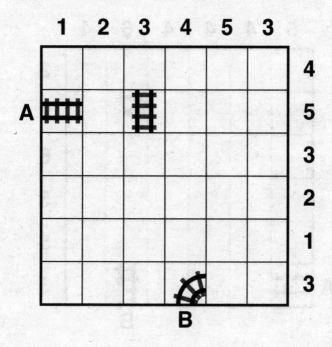

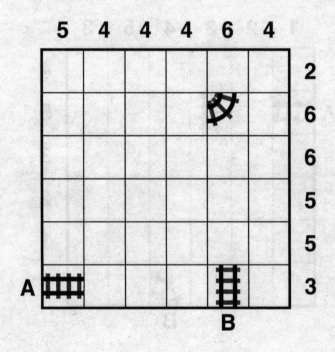

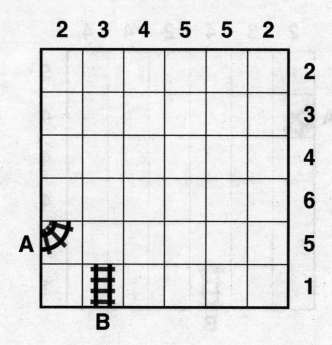

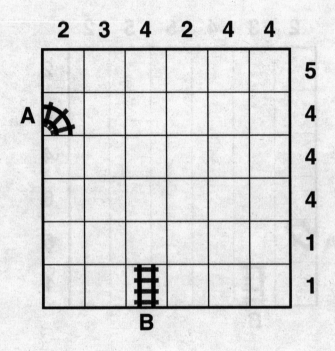

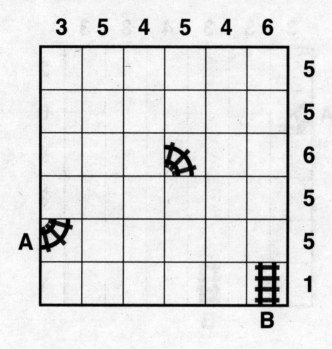

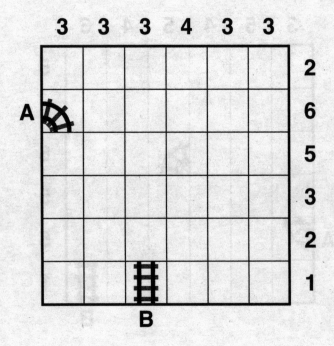

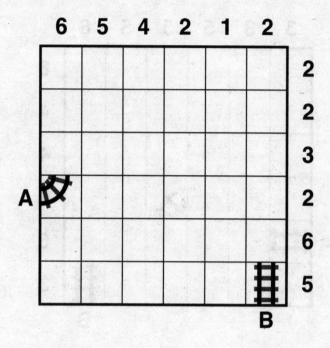

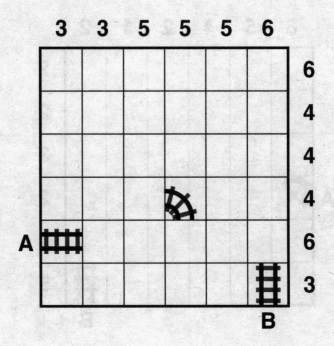

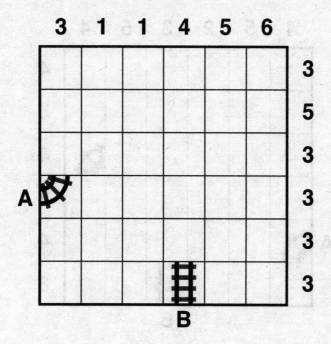

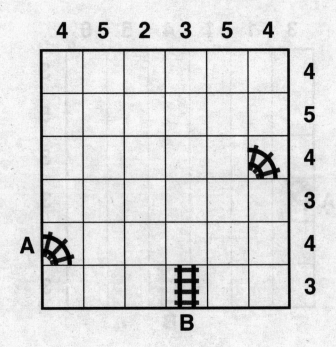

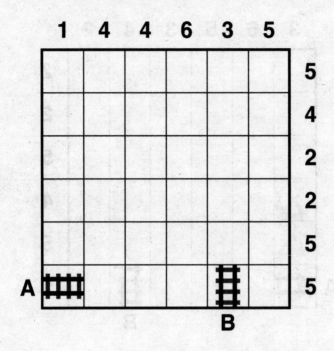

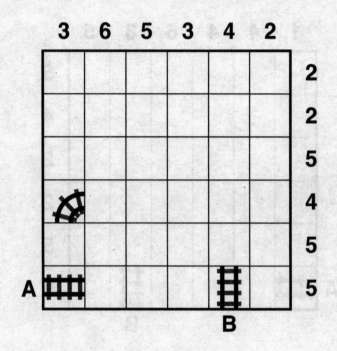

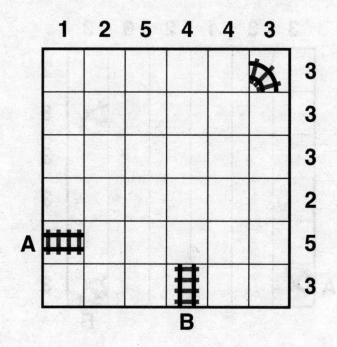

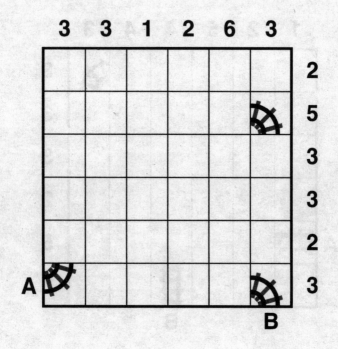

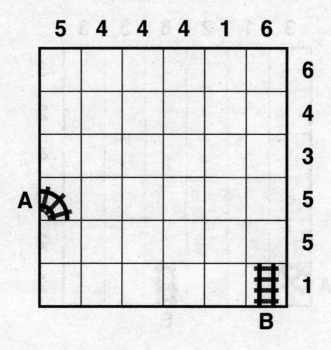

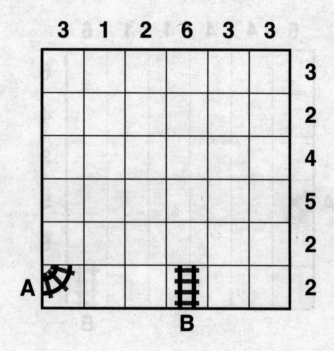

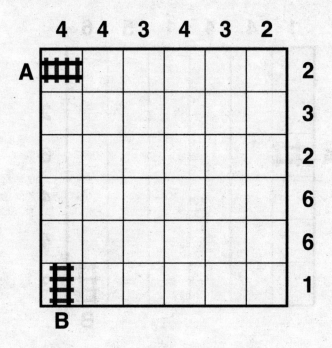

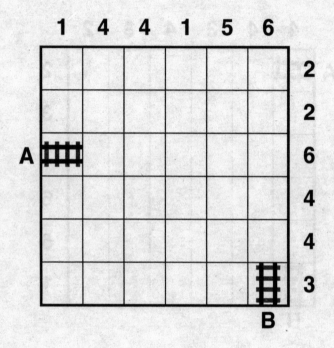

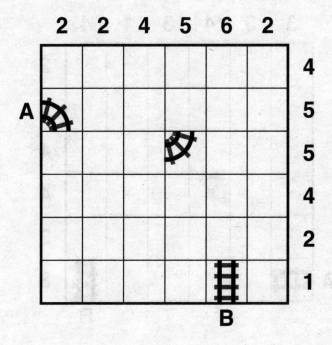

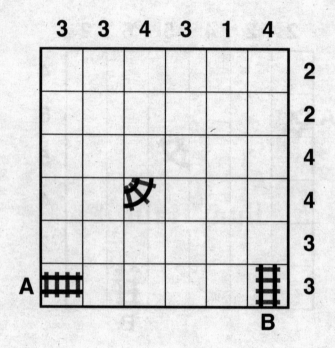

Medium Train Tracks

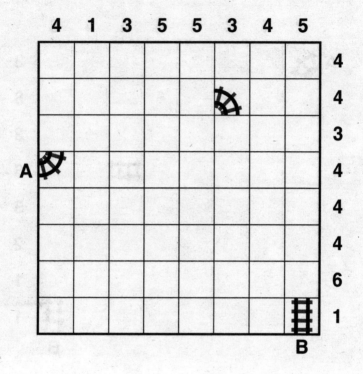

31

4 1 3 5 5 3 4 5

4
4
3
4
4
4
6
1

A

B

Medium

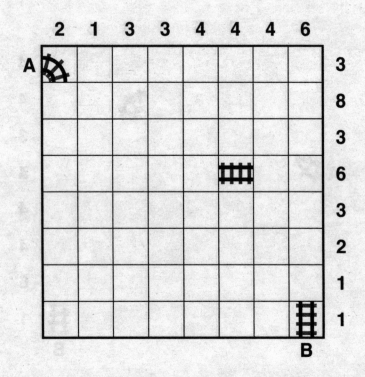

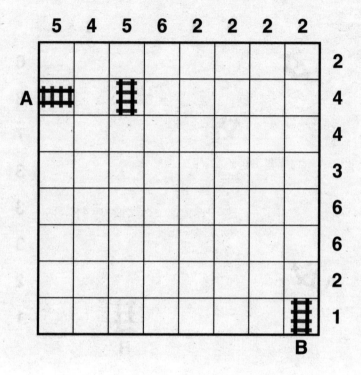

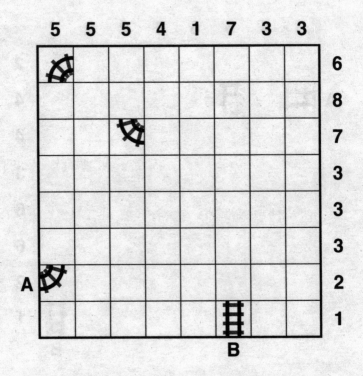

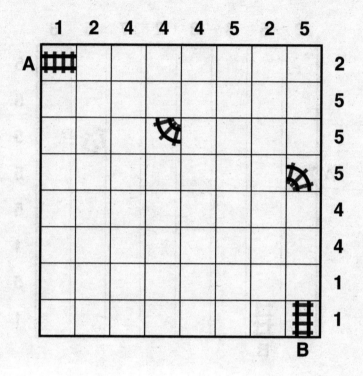

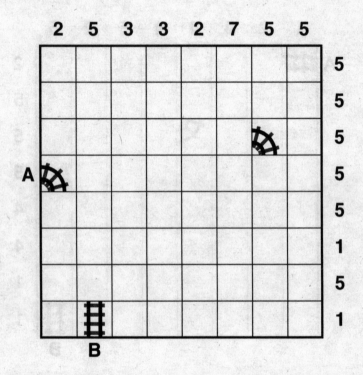

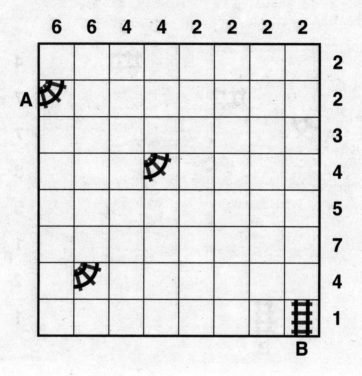

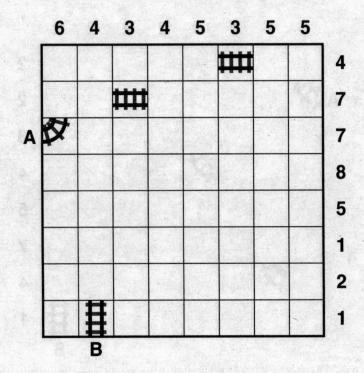

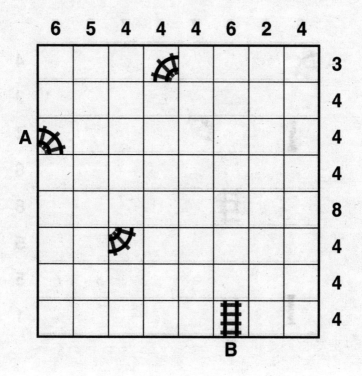

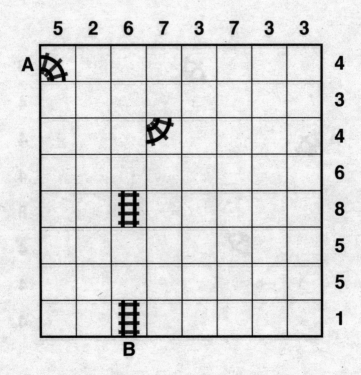

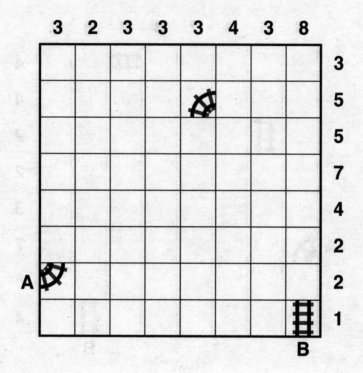

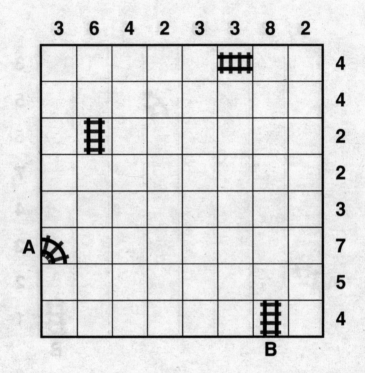

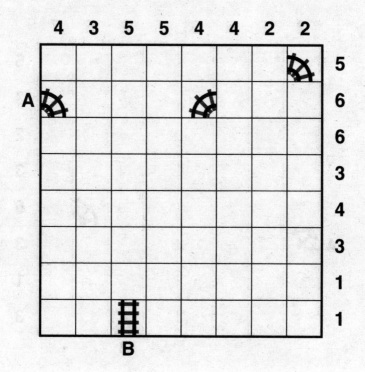

43

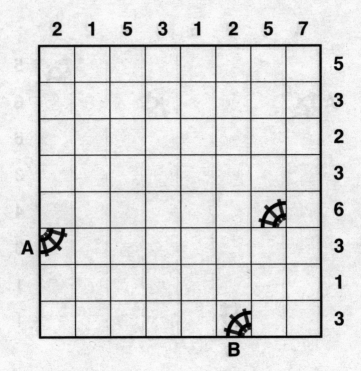

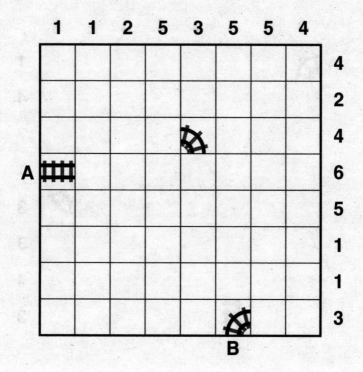

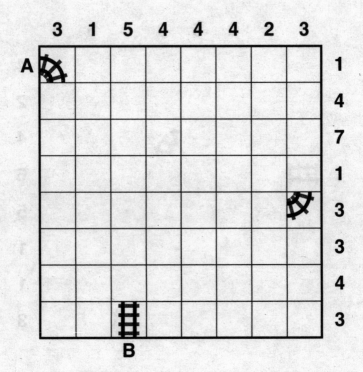

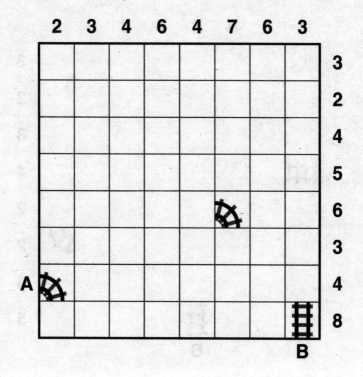

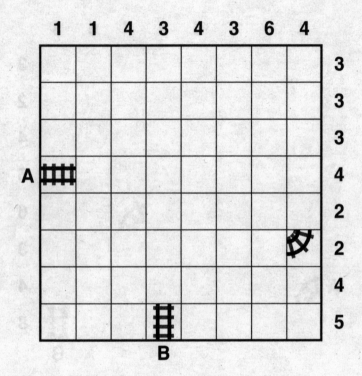

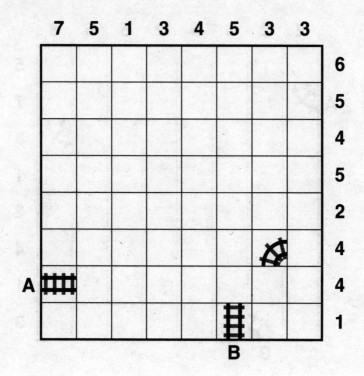

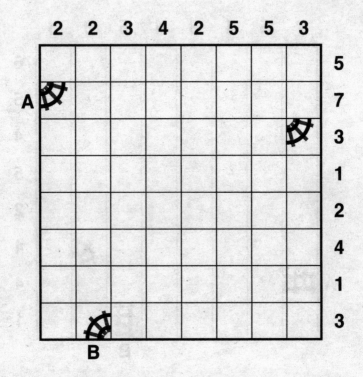

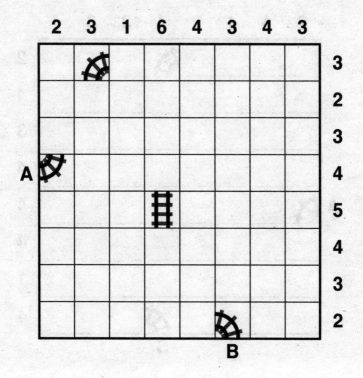

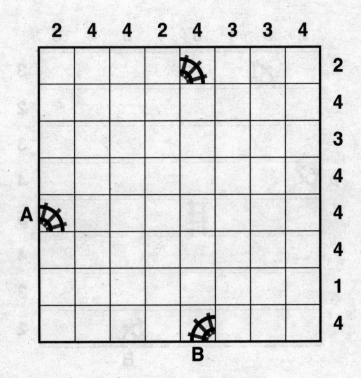

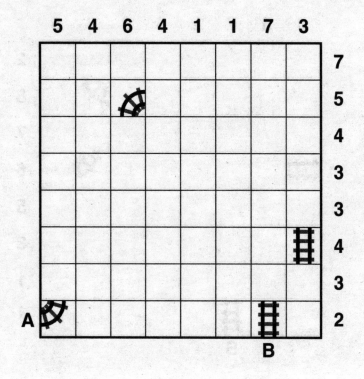

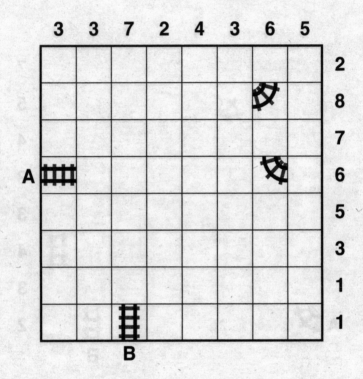

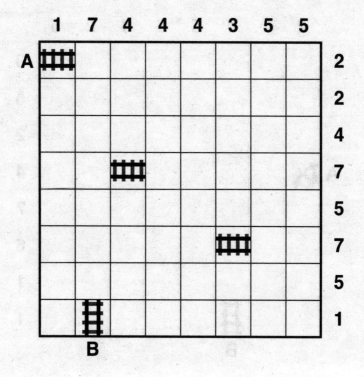

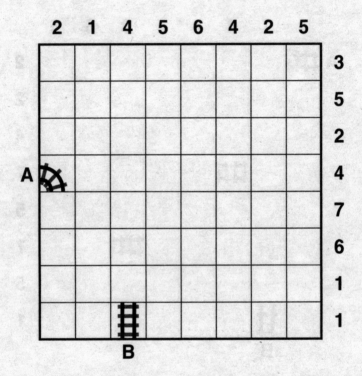

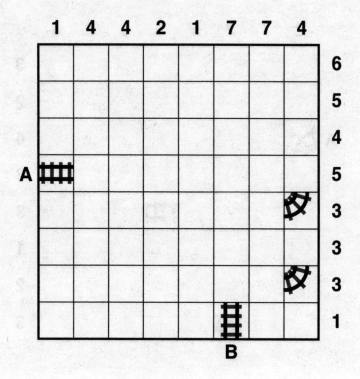

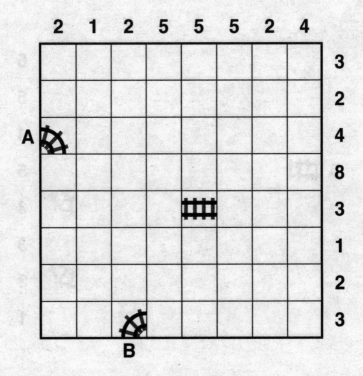

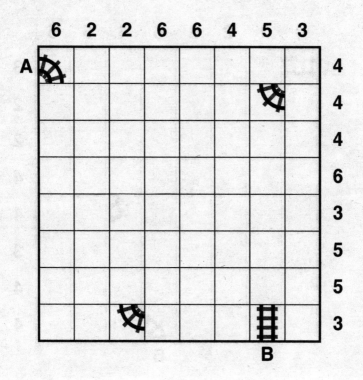

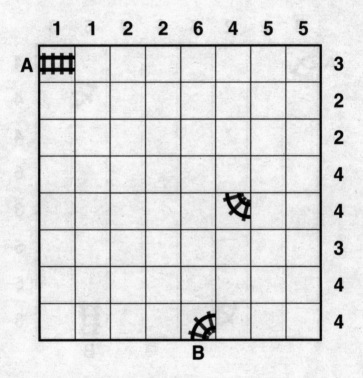

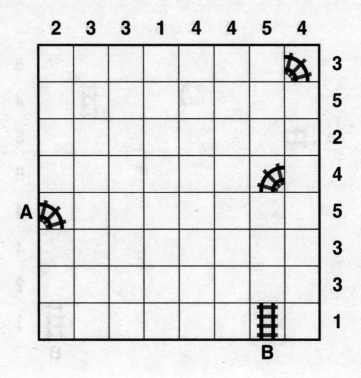

Medium

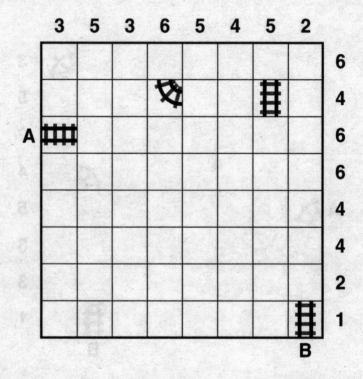

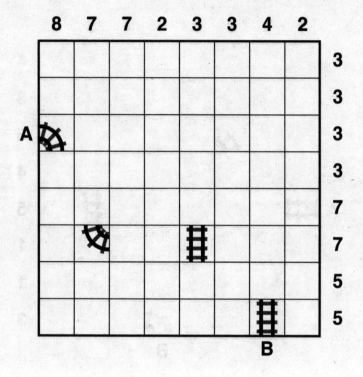

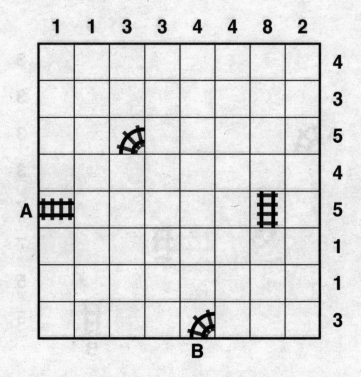

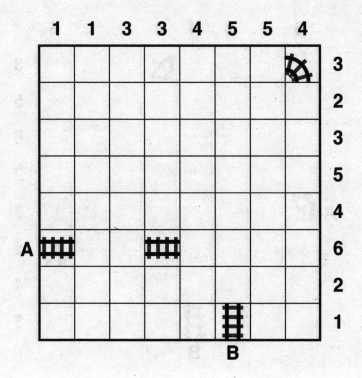

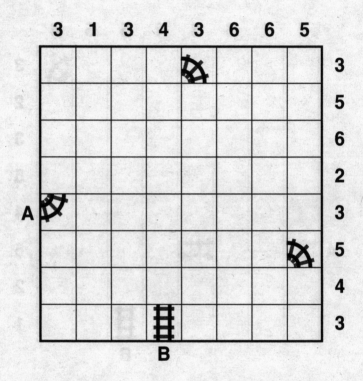

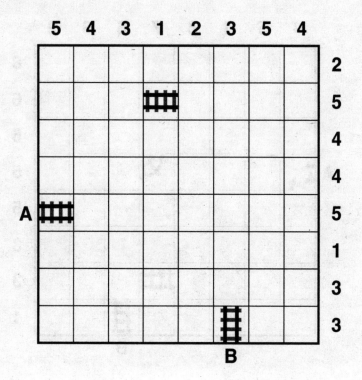

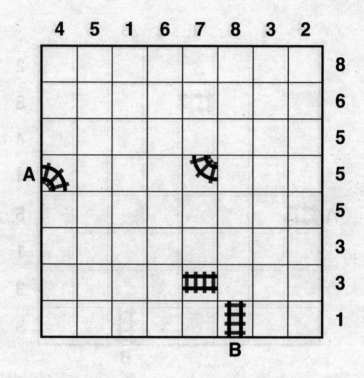

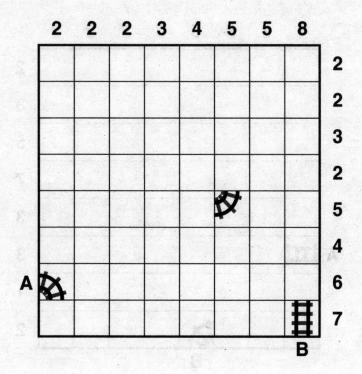

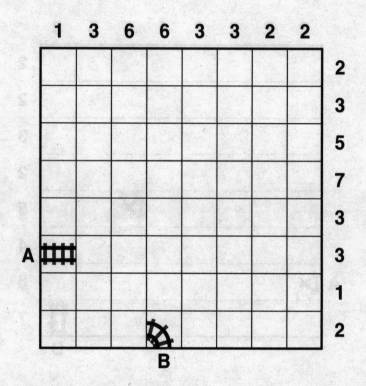

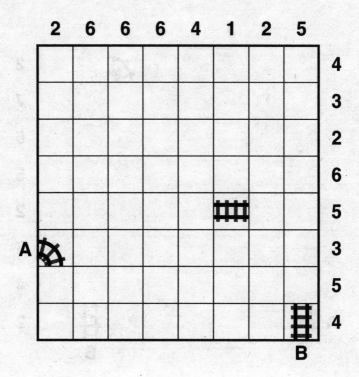

Medium

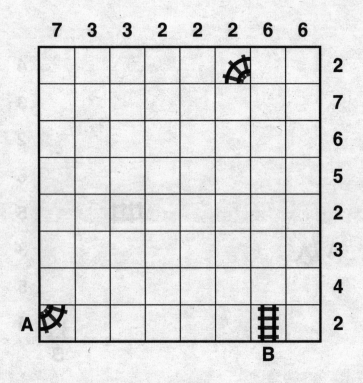

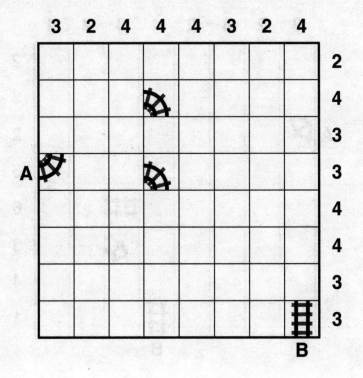

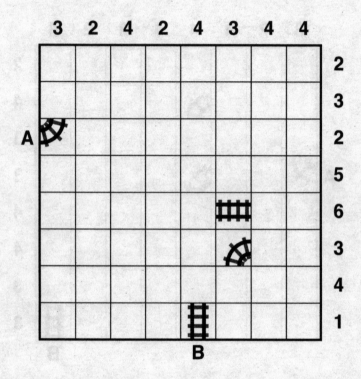

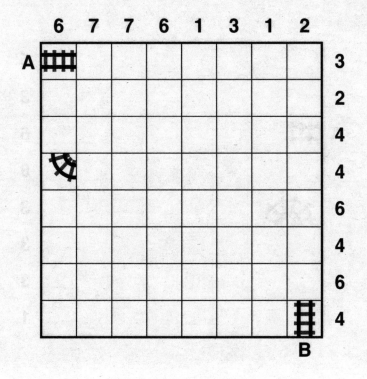

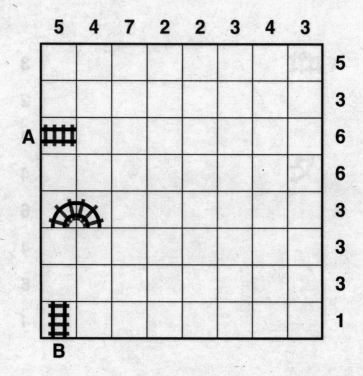

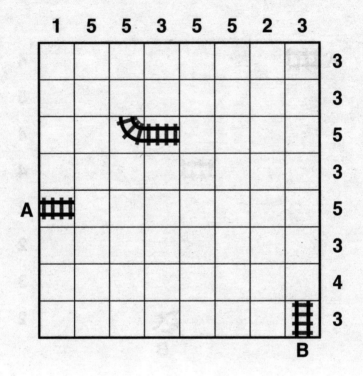

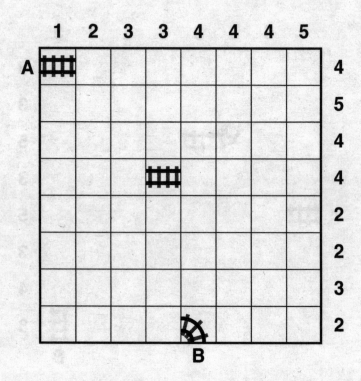

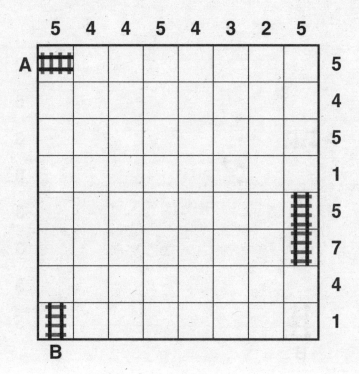

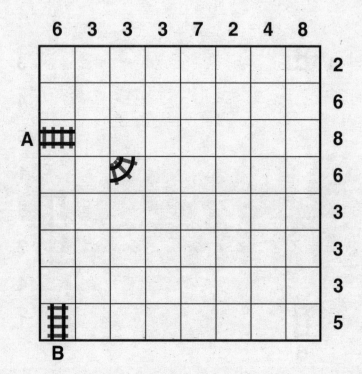

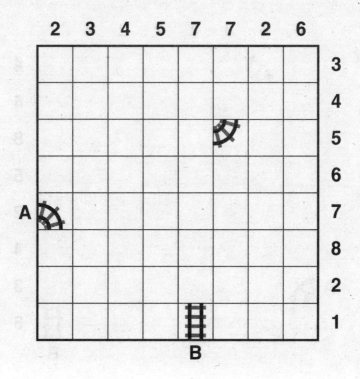

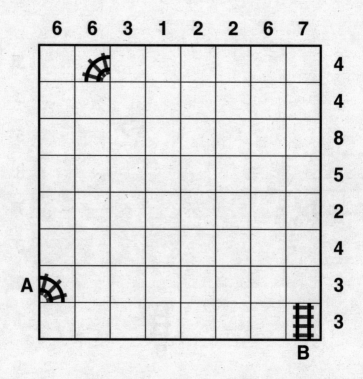

83

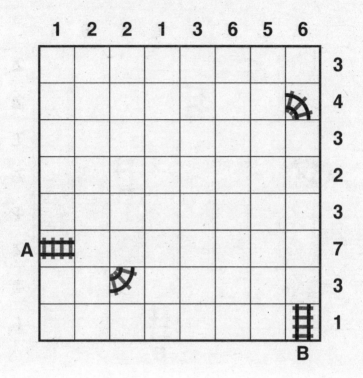

Medium

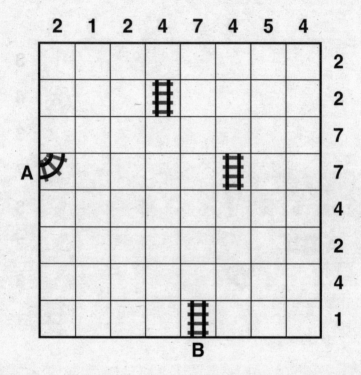

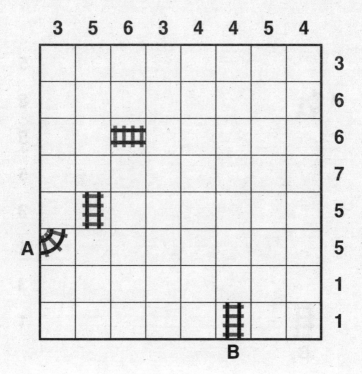

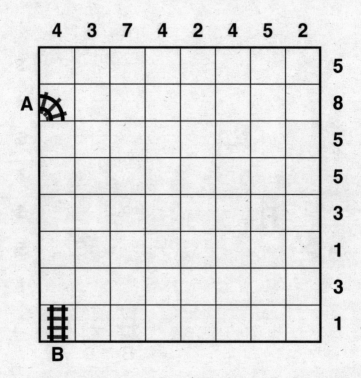

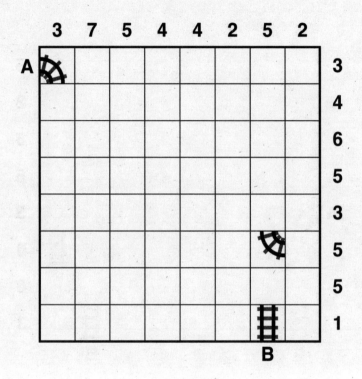

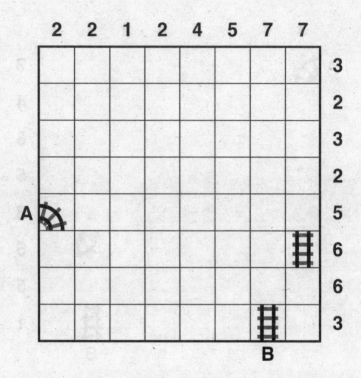

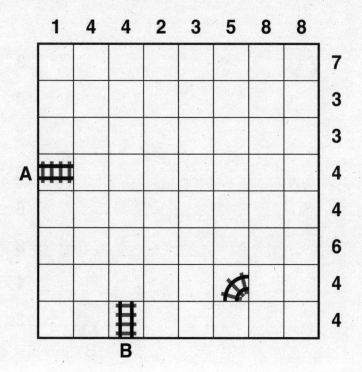

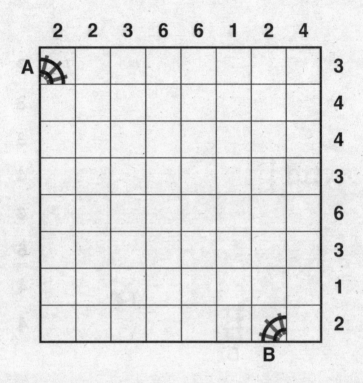

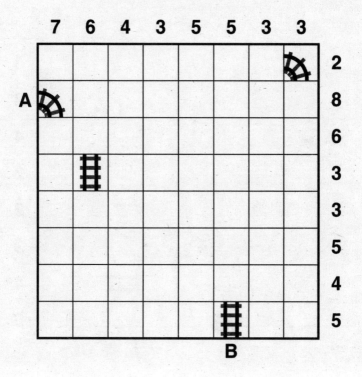

Medium

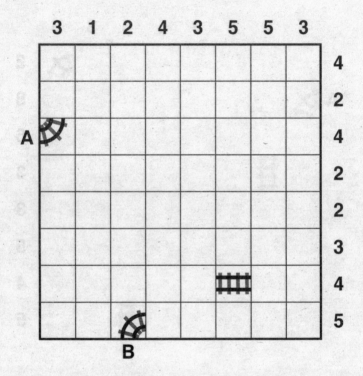

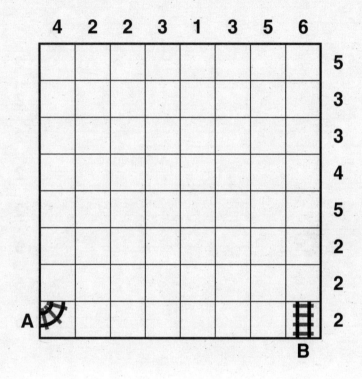

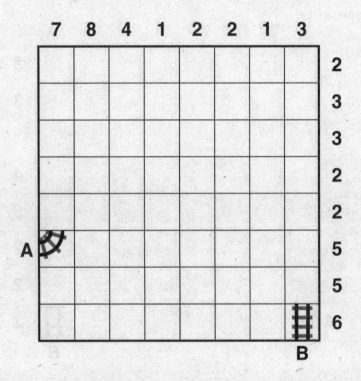

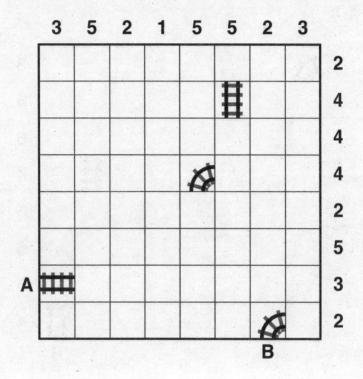

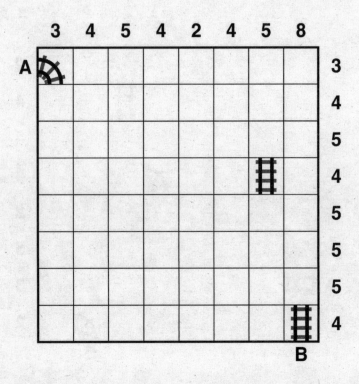

The Times Train Tracks

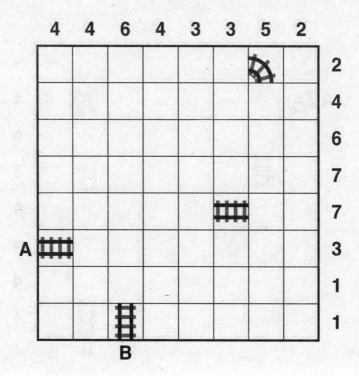

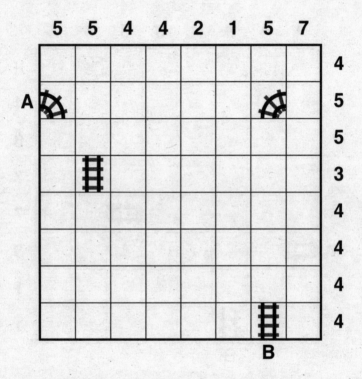

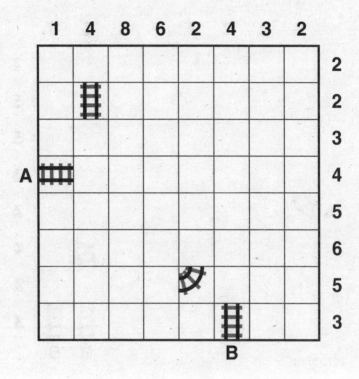

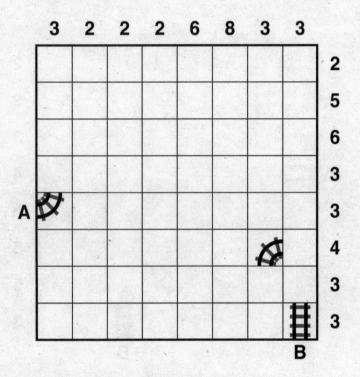

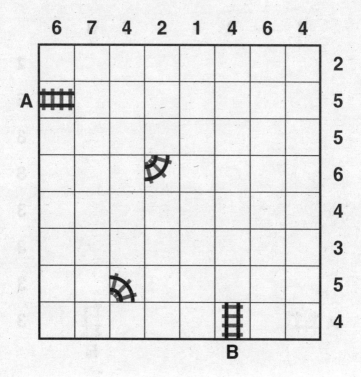

Medium

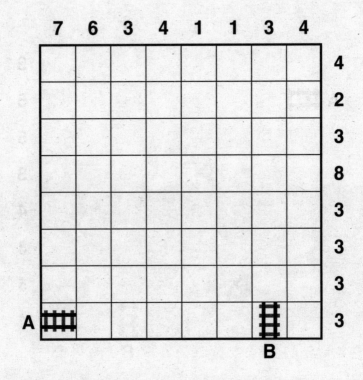

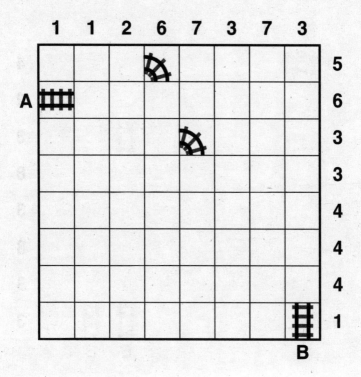

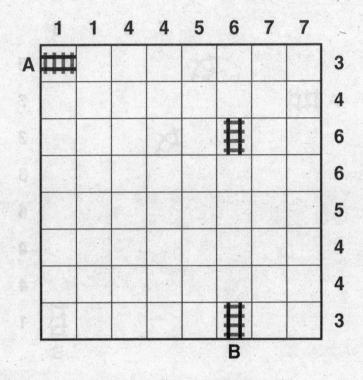

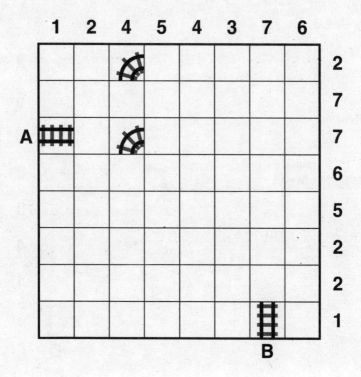

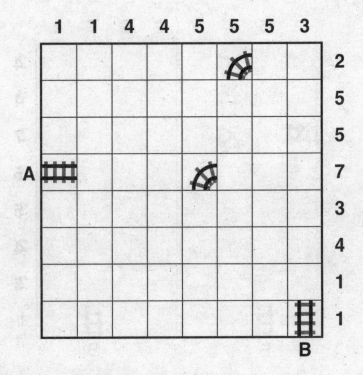

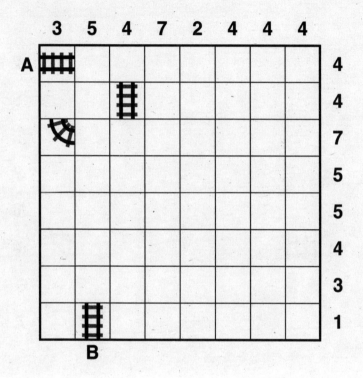

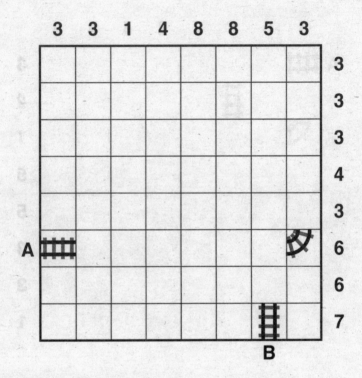

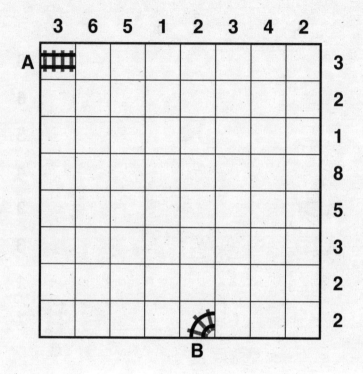

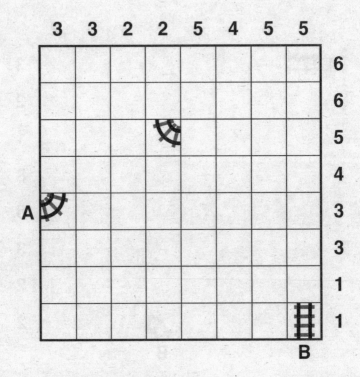

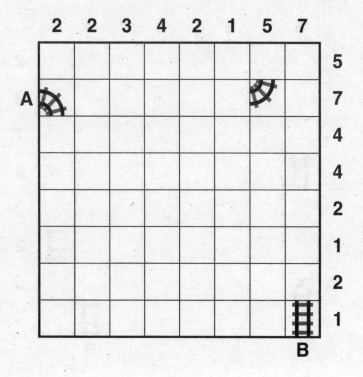

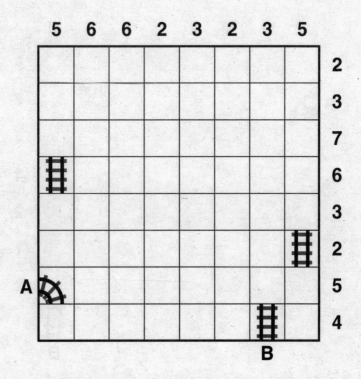

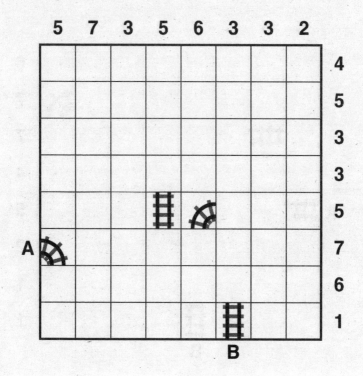

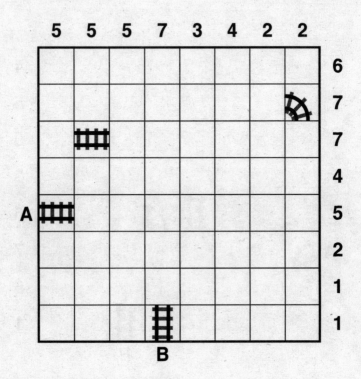

The Times Train Tracks

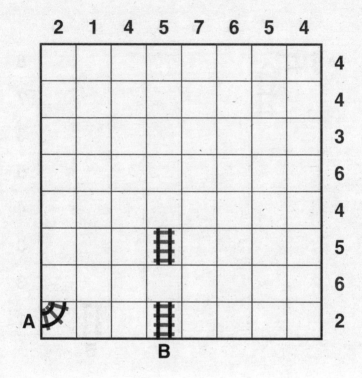

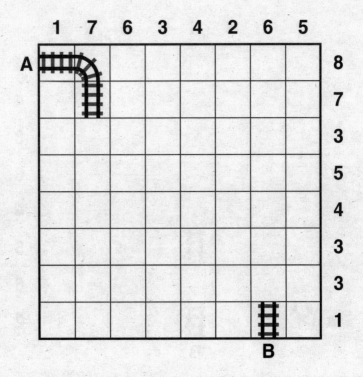

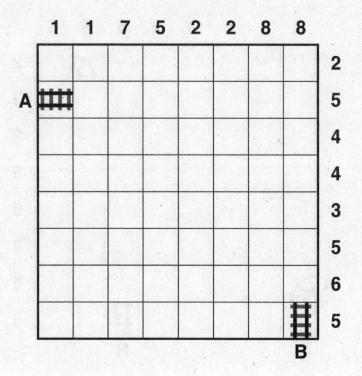

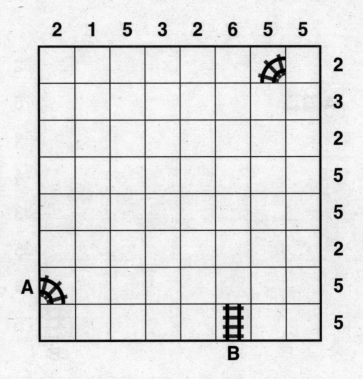

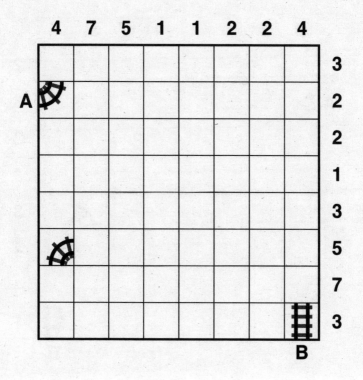

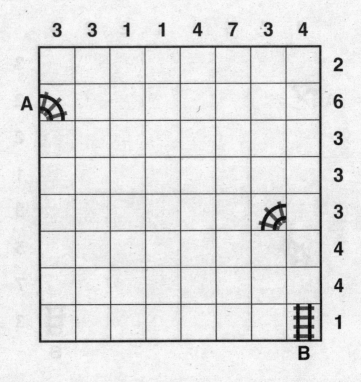

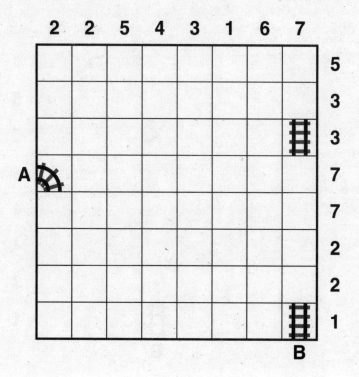

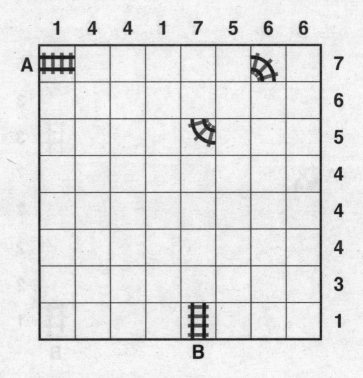

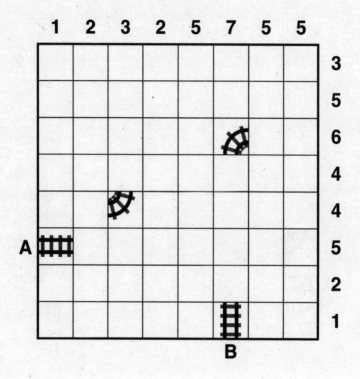

Medium

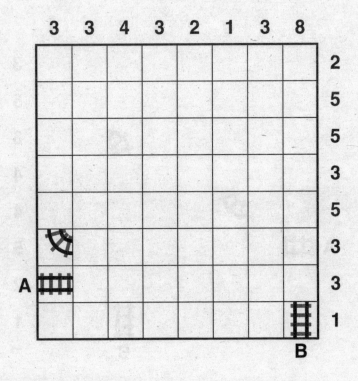

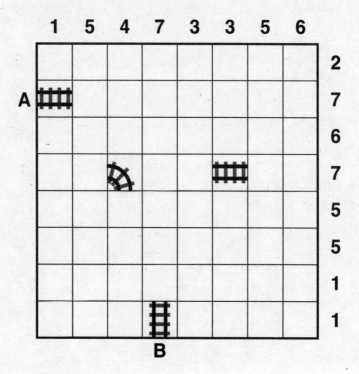

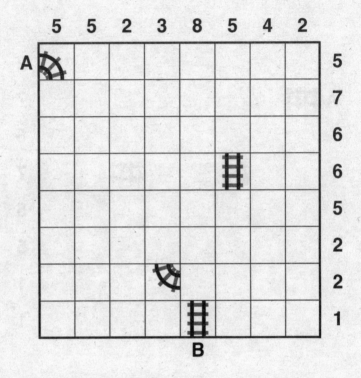

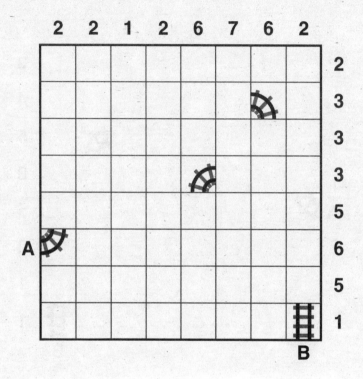

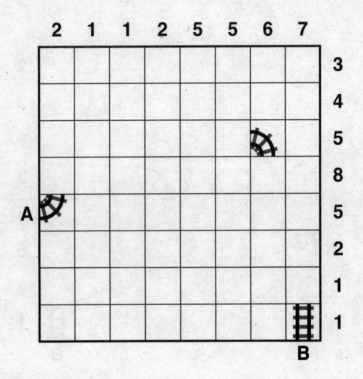

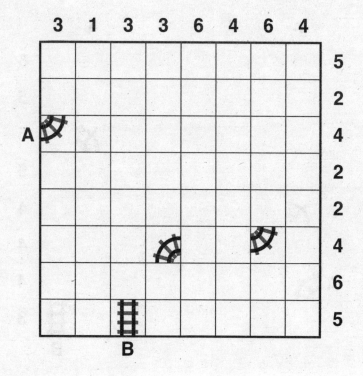

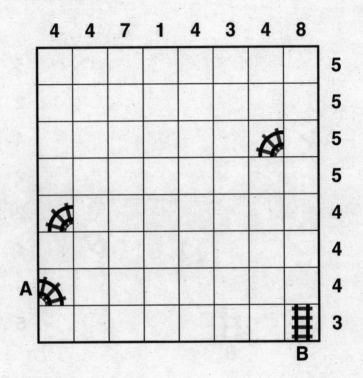

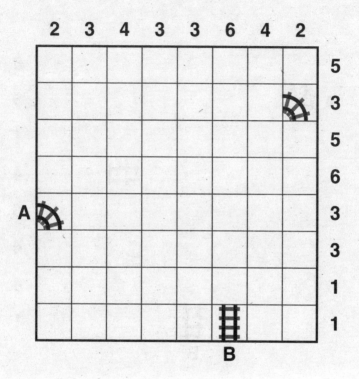

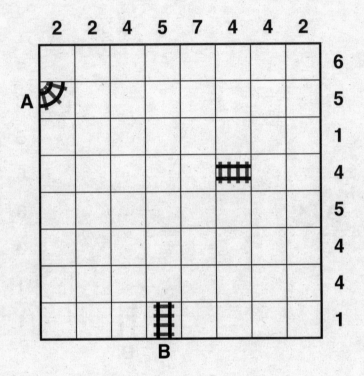

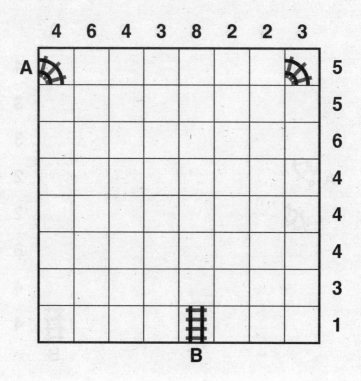

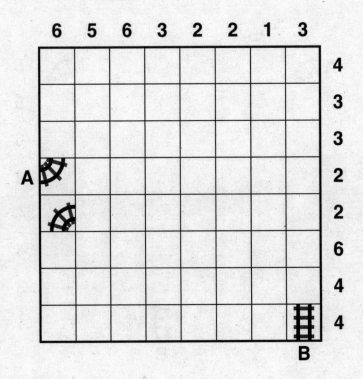

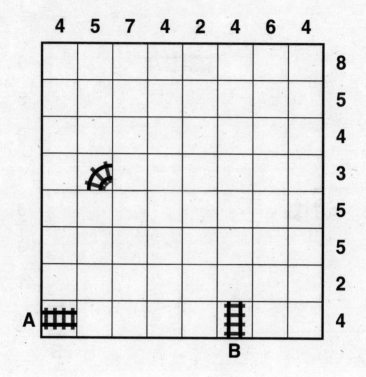

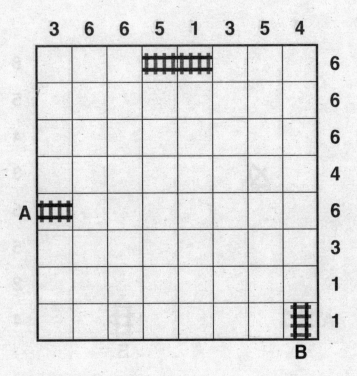

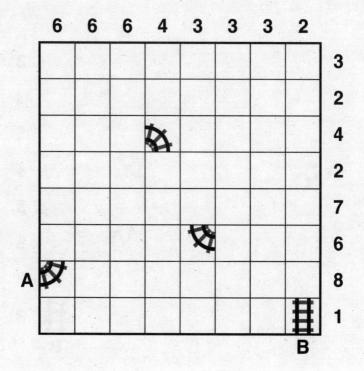

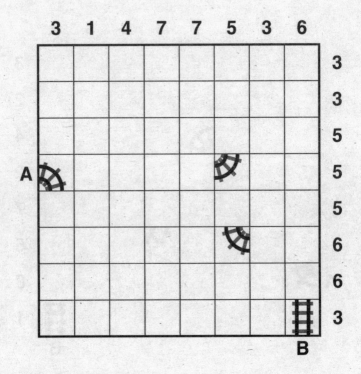

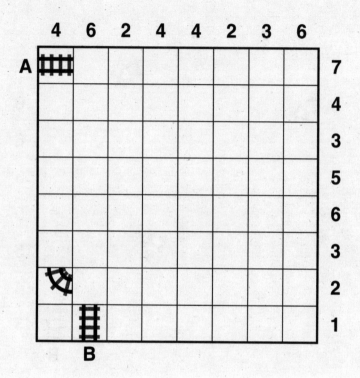

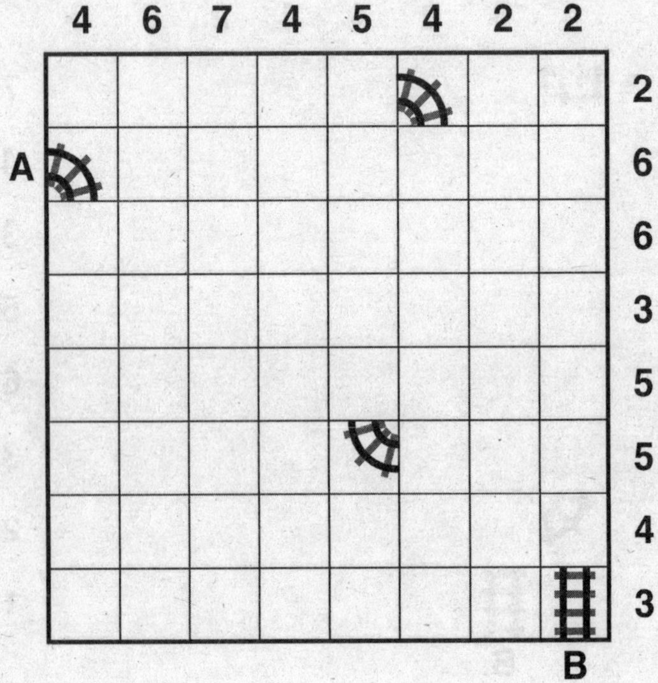
4 6 7 4 5 4 2 2
2
6
6
3
5
5
4
3
A
B

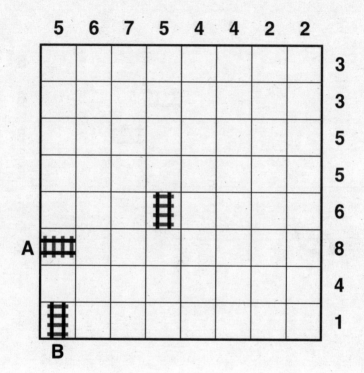

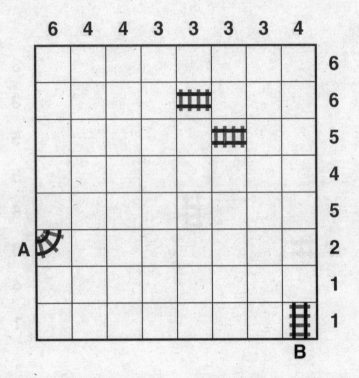

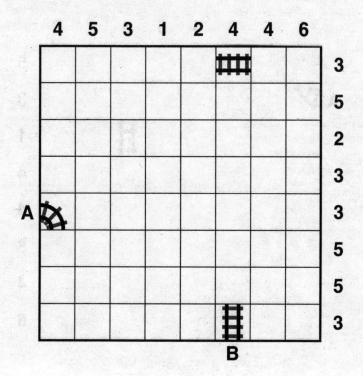

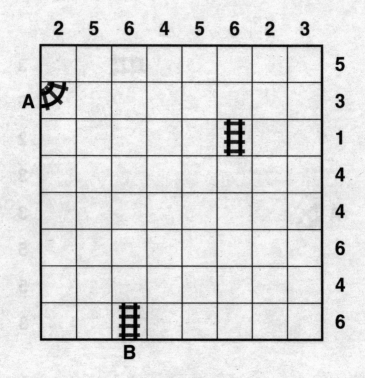

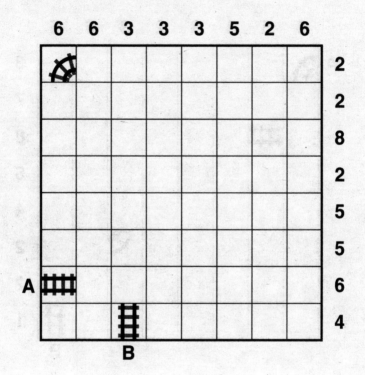

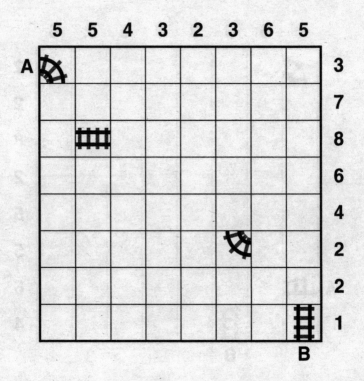

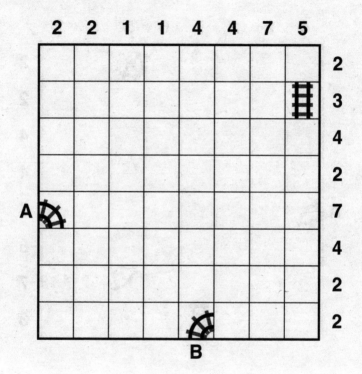

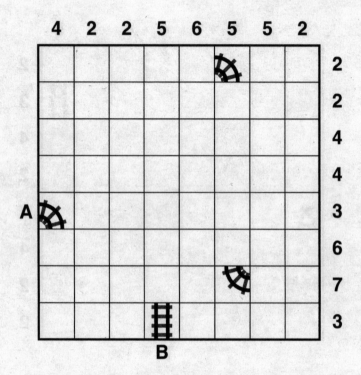

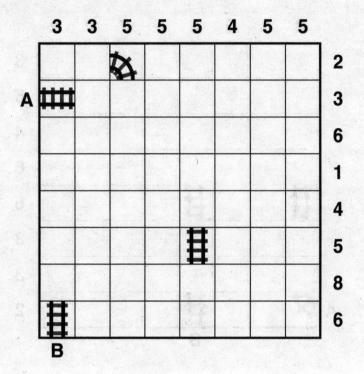

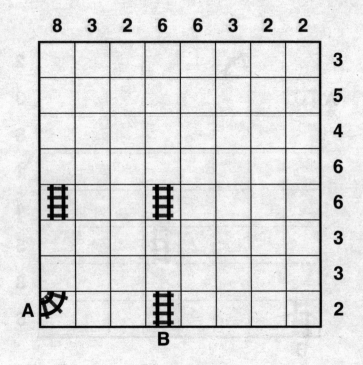

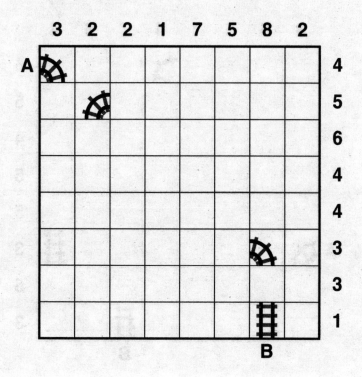

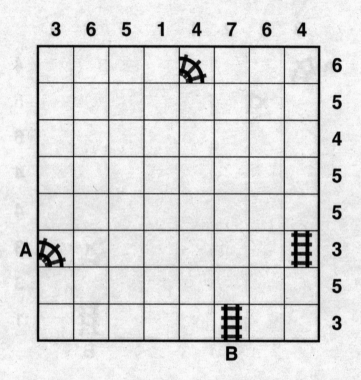

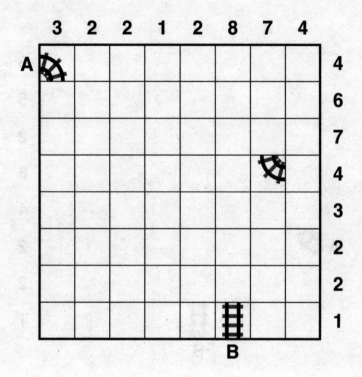

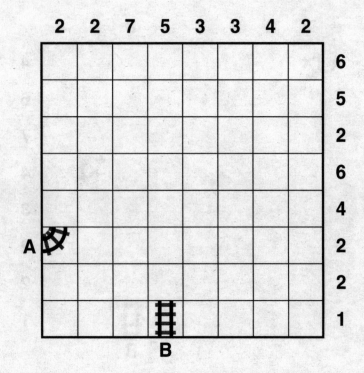

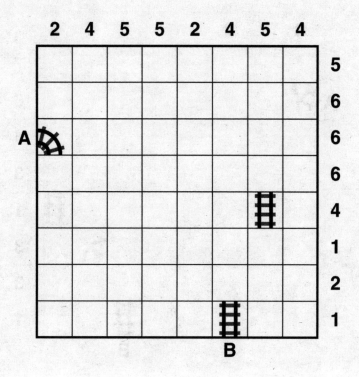

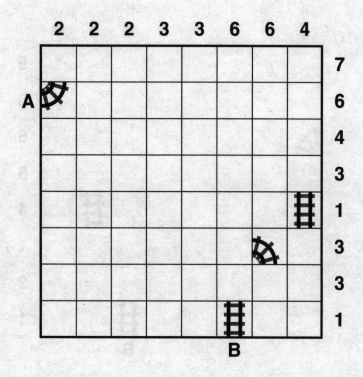

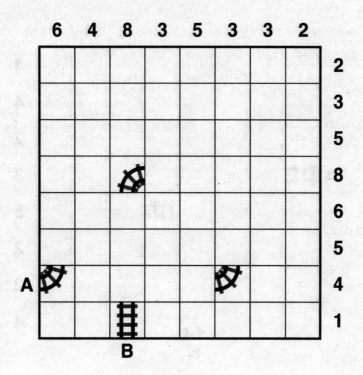

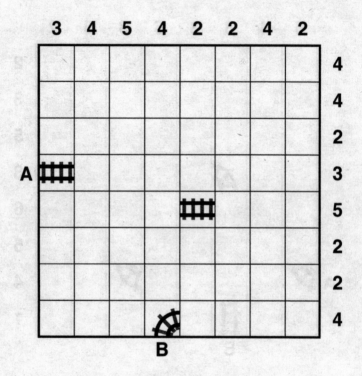

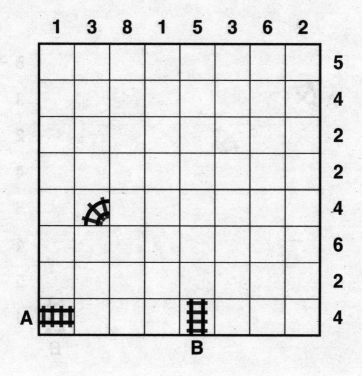

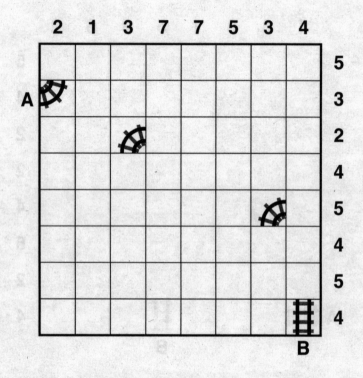

Hard Train Tracks

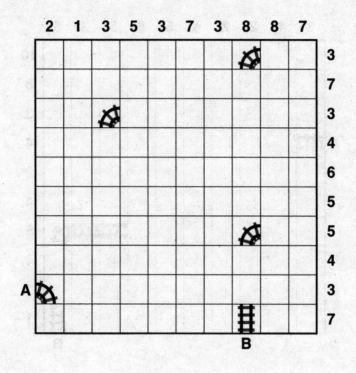

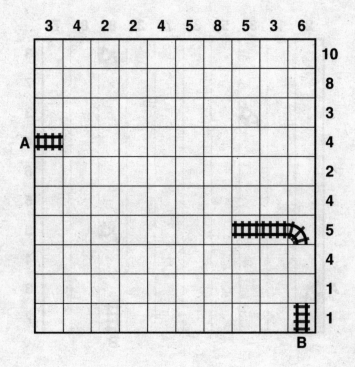

The Times Train Tracks

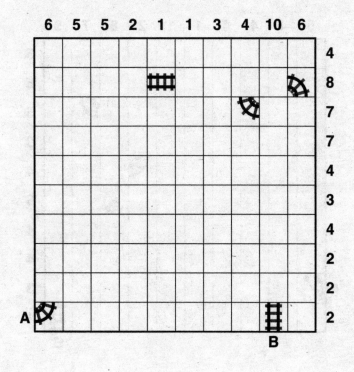

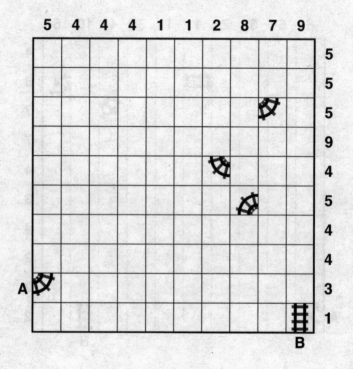

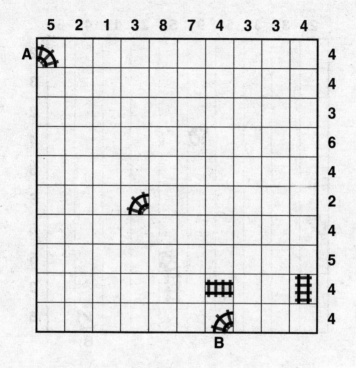

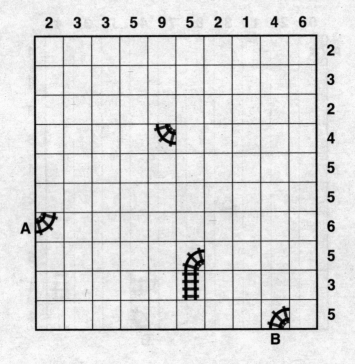

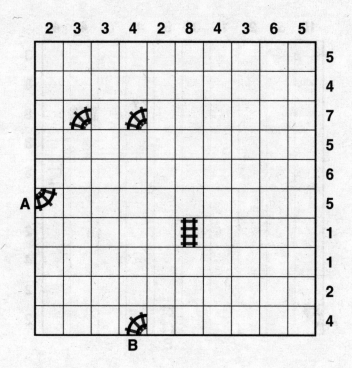

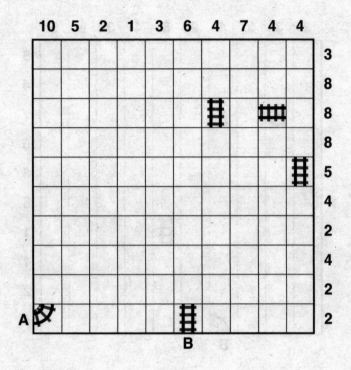

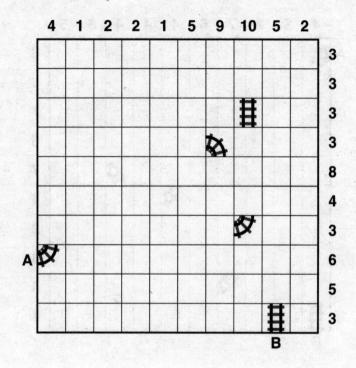

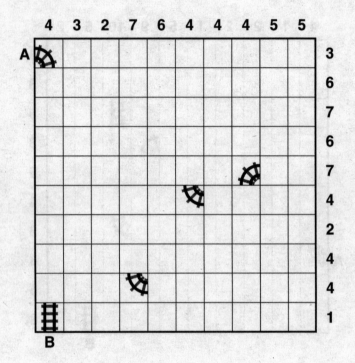

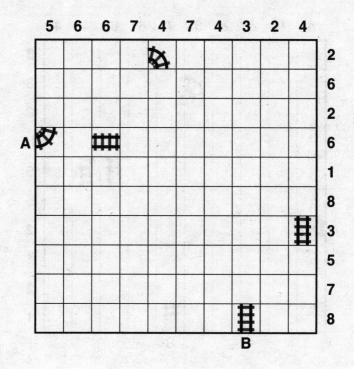

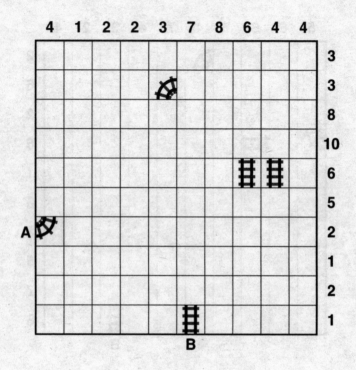

172

The Times Train Tracks

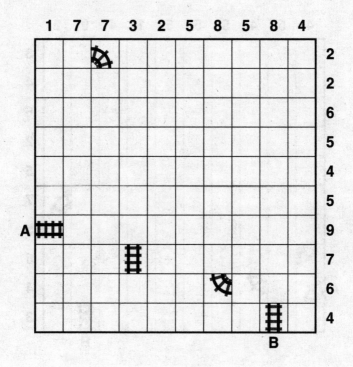

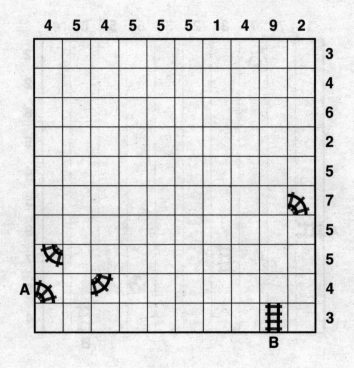

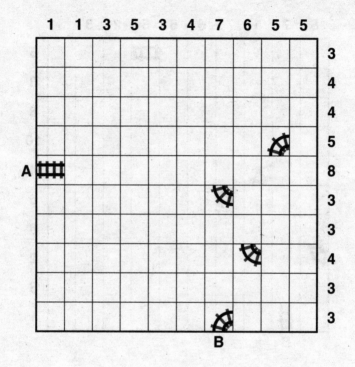

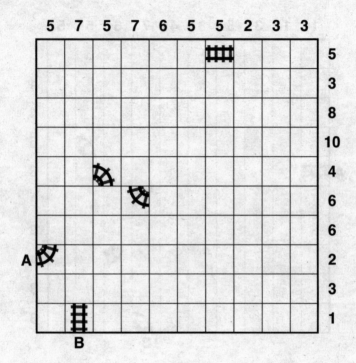

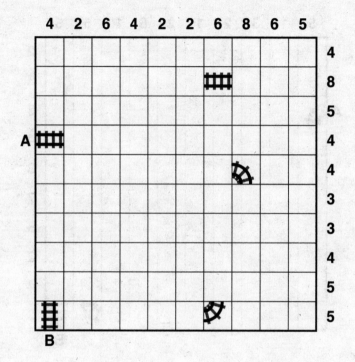

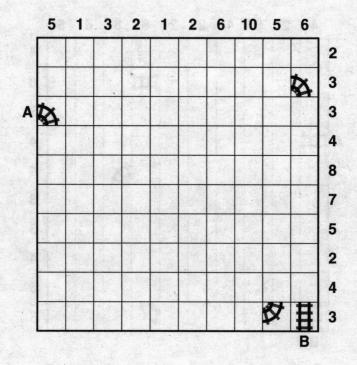

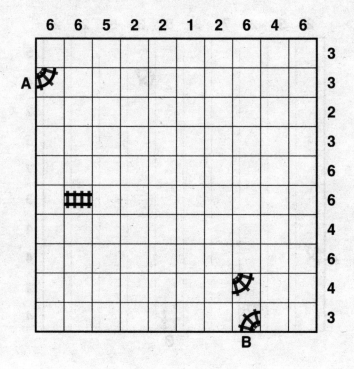

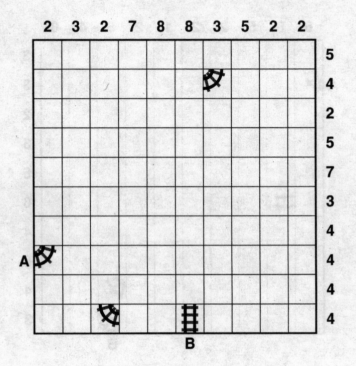

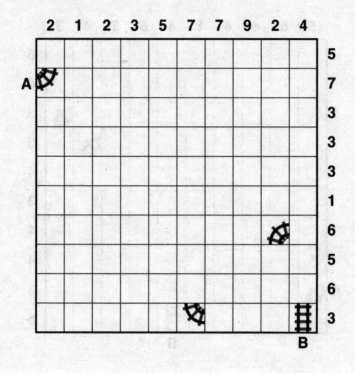

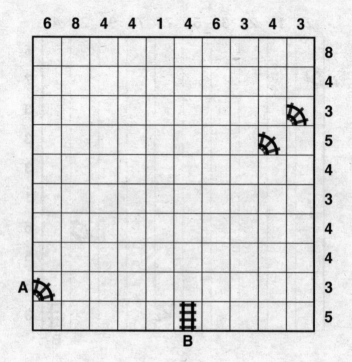

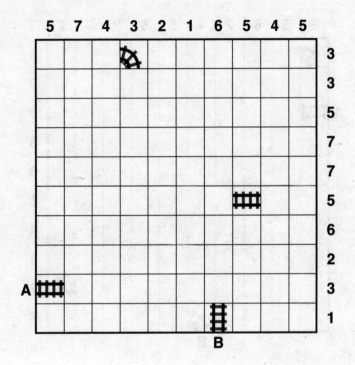

5 7 4 3 2 1 6 5 4 5

3
3
5
7
7
5
6
2
3
1

A
B

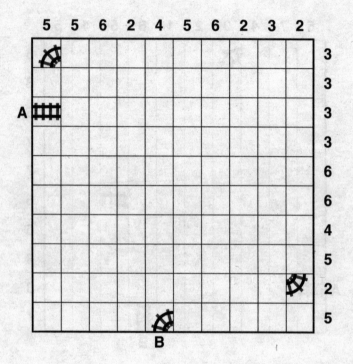

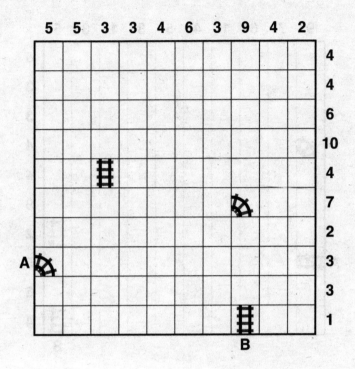

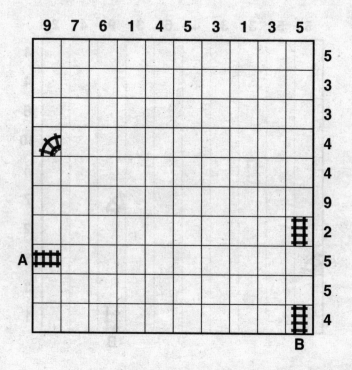

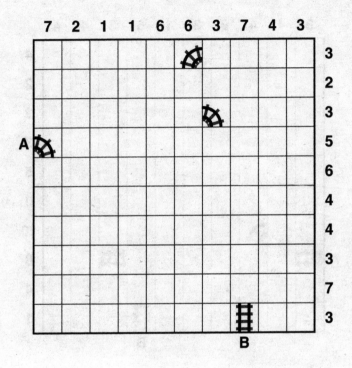

187

Hard

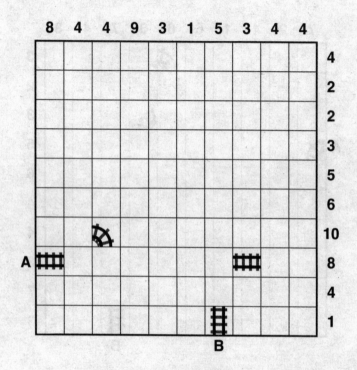

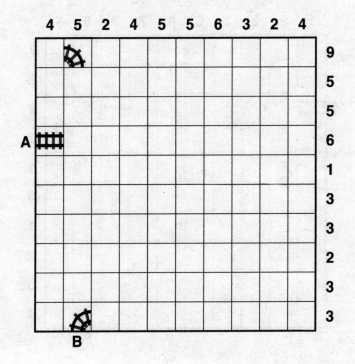

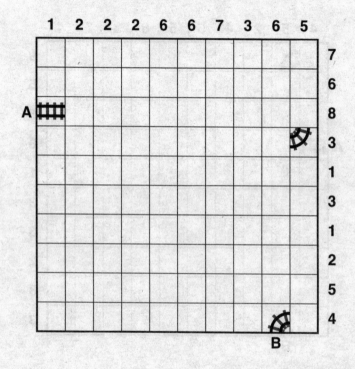

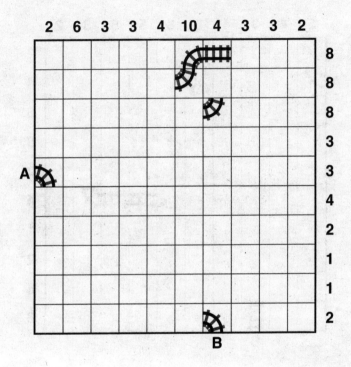

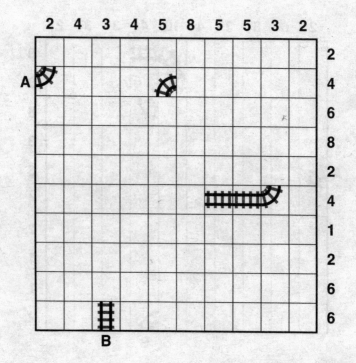

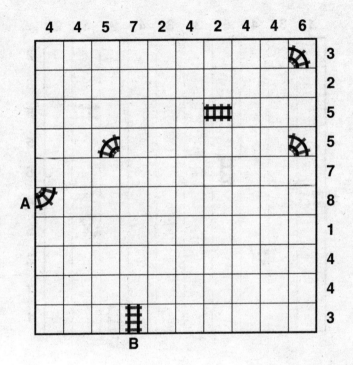

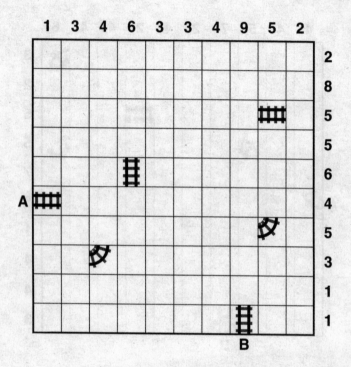

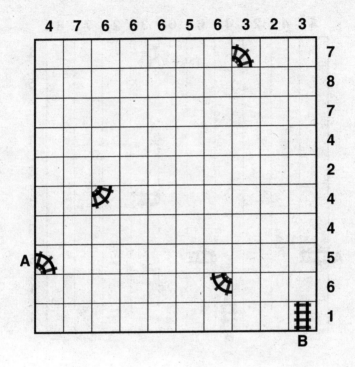

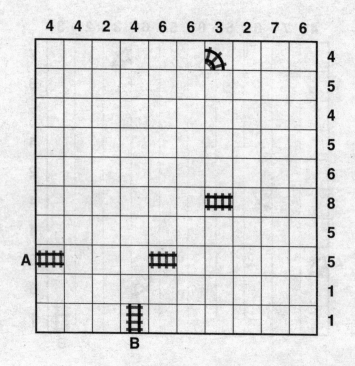

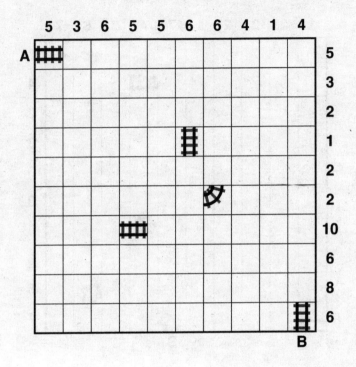

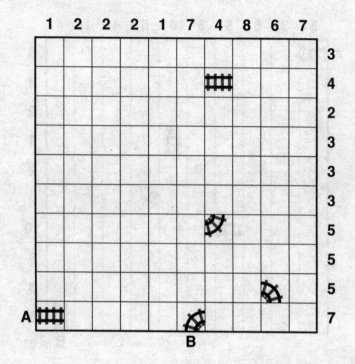

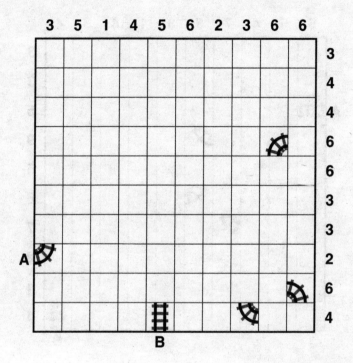

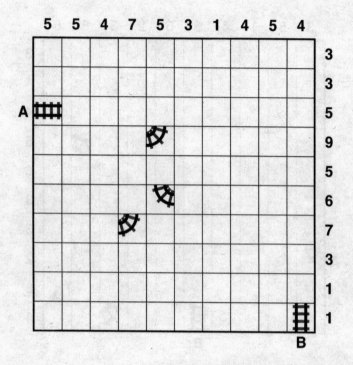

Solutions

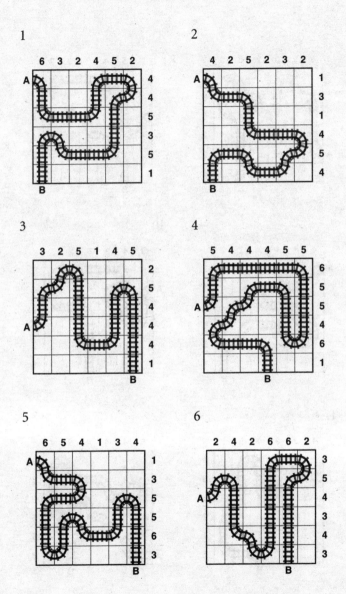

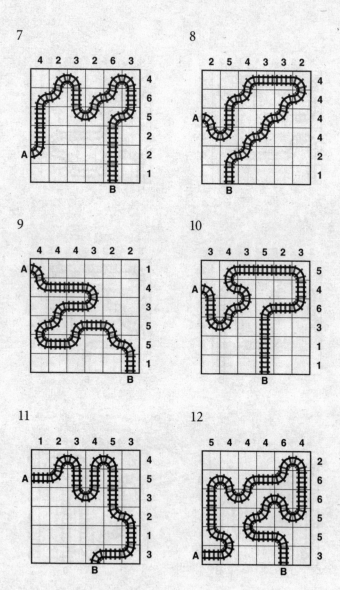

The Times Train Tracks

13

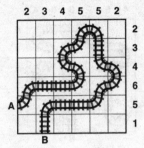

14

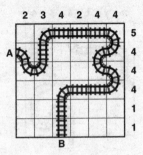

15

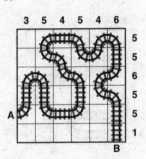

16

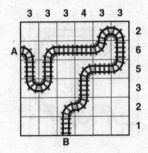

17

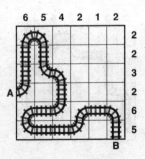

18

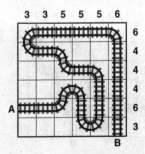

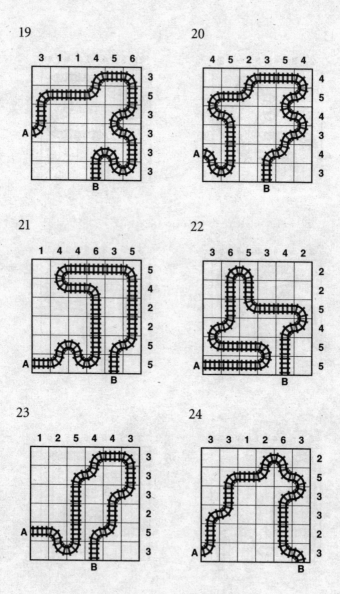

19

20

21

22

23

24

The Times Train Tracks

25

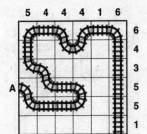

26

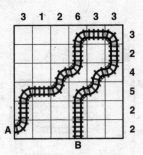

27

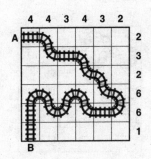

28

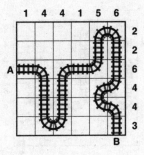

29

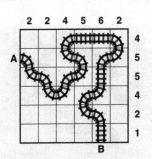

30

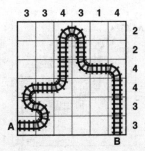

31

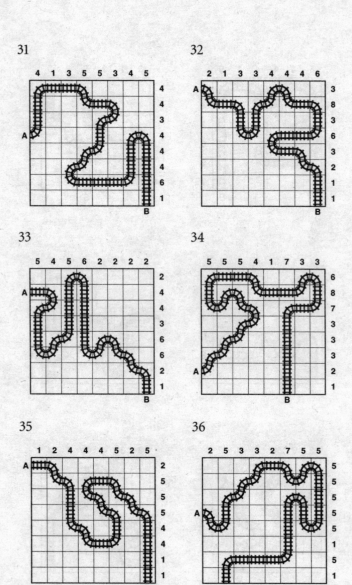

32

33

34

35

36

The Times Train Tracks

37

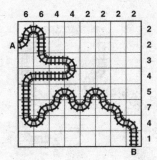

38

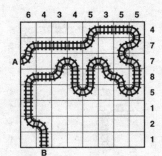

39

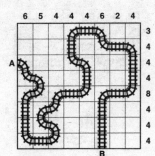

40

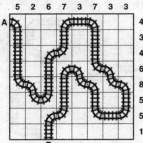

41

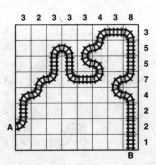

42

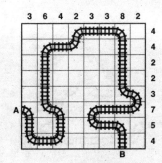

Solutions

43

44

45

46

47

48

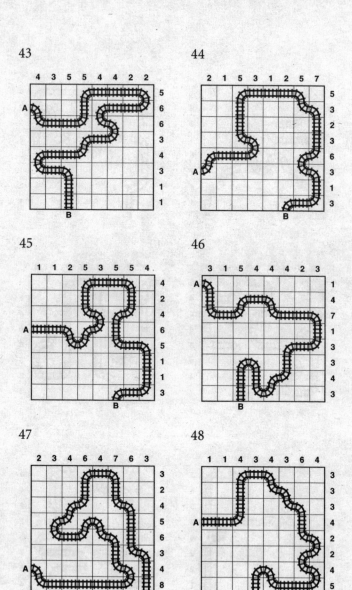

The Times Train Tracks

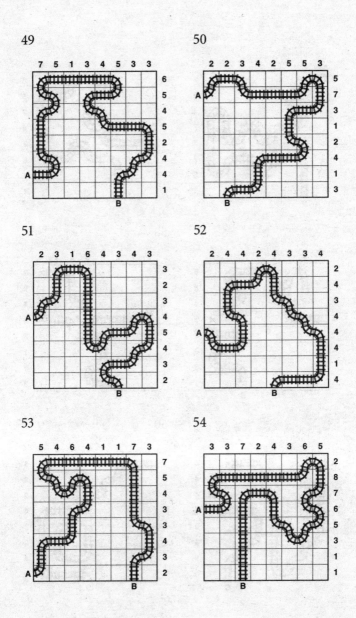

49

50

51

52

53

54

Solutions

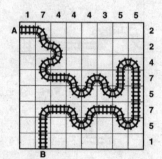

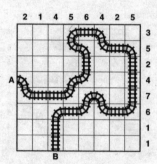

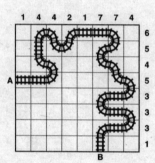

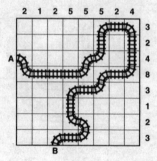

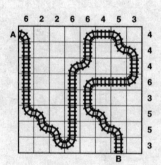

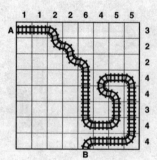

The Times Train Tracks

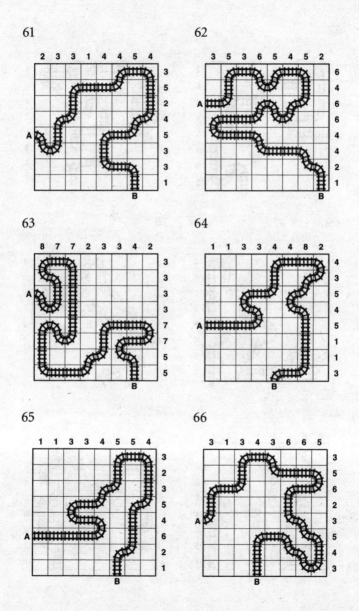

67

68

69

70

71

72

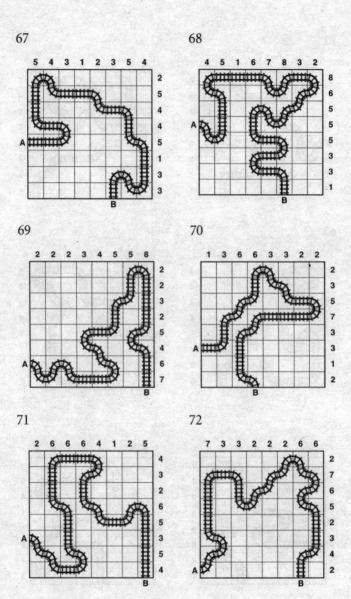

73

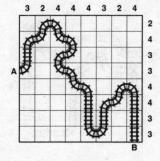

74

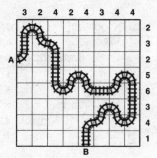

75

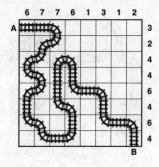

76

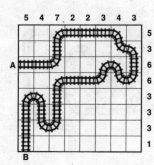

77

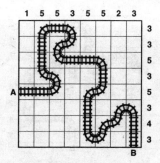

78

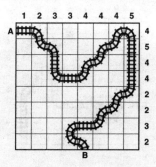

Solutions

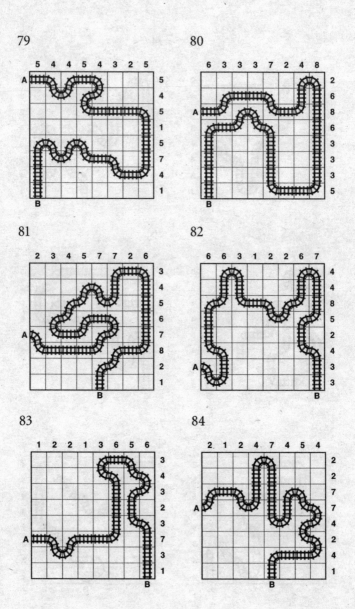

79

80

81

82

83

84

The Times Train Tracks

85

3 5 6 3 4 4 5 4

3
6
6
7
5
5
1
1

A

B

86

4 3 7 4 2 4 5 2

5
8
5
5
3
1
3
1

A

B

87

3 7 5 4 4 2 5 2

3
4
6
5
3
5
5
1

A

B

88

2 2 1 2 4 5 7 7

3
2
3
2
5
6
6
3

A

B

89

1 4 4 2 3 5 8 8

7
3
3
4
4
6
4
4

A

B

90

2 2 3 6 6 1 2 4

3
4
4
3
6
3
1
2

A

B

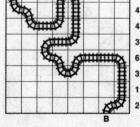

Solutions

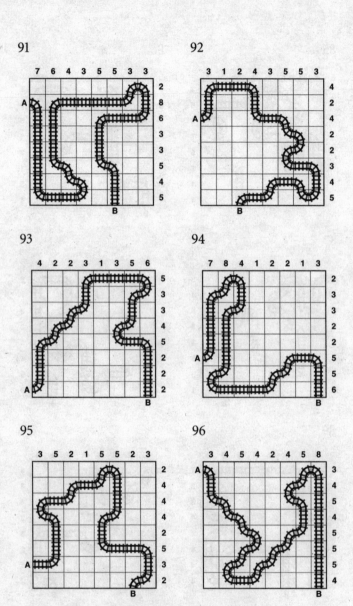

The Times Train Tracks

97

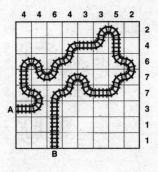

98

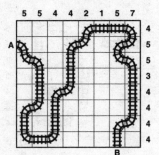

99

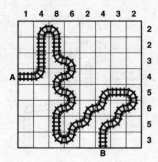

100

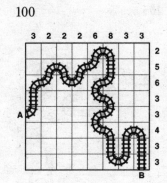

101

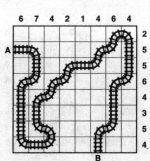

102

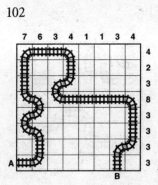

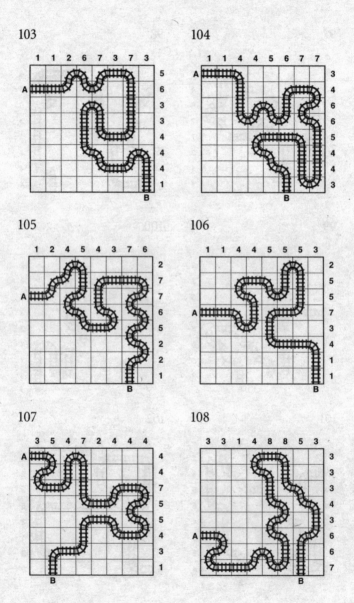

The Times Train Tracks

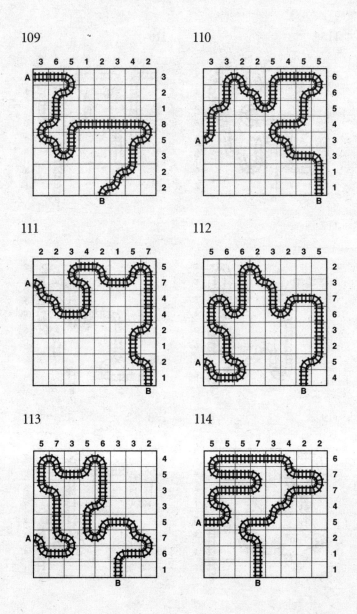

109

110

111

112

113

114

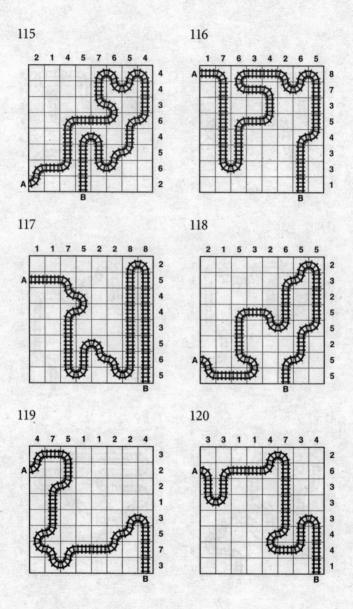

The Times Train Tracks

121

2	2	5	4	3	1	6	7	
								5
								3
								3
A									7
								7
								2
								2
								1
B

122

| 1 | 4 | 4 | 1 | 7 | 5 | 6 | 6 |
A
7 6 5 4 4 3 1
B

123

| 1 | 2 | 3 | 2 | 5 | 7 | 5 | 5 |
3 5 6 4 4 5 2 1
A
B

124

| 3 | 3 | 4 | 3 | 2 | 1 | 3 | 8 |
2 5 5 3 5 3 3 1
A
B

125

| 1 | 5 | 4 | 7 | 3 | 3 | 5 | 6 |
A
2 7 6 7 5 1 1
B

126

| 5 | 5 | 2 | 3 | 8 | 5 | 4 | 2 |
A
5 7 6 5 2 2 1
B

Solutions

127

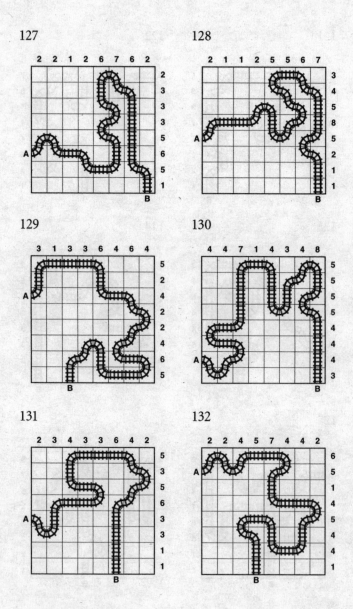

128

129

130

131

132

The Times Train Tracks

133

| 4 | 6 | 4 | 3 | 8 | 2 | 2 | 3 |

								5
								5
								6
								4
								4
								4
								3
								1

134

| 6 | 5 | 6 | 3 | 2 | 2 | 1 | 3 |

								4
								3
								3
								2
								2
								6
								4
								4

135

| 4 | 5 | 7 | 4 | 2 | 4 | 6 | 4 |

								8
								5
								4
								3
								5
								5
								2
								4

136

| 3 | 6 | 6 | 5 | 1 | 3 | 5 | 4 |

								6
								6
								6
								4
								6
								3
								1
								1

137

| 6 | 6 | 6 | 4 | 3 | 3 | 3 | 2 |

								3
								2
								4
								2
								7
								6
								8
								1

138

| 3 | 1 | 4 | 7 | 7 | 5 | 3 | 6 |

								3
								3
								5
								5
								5
								6
								6
								3

Solutions

139

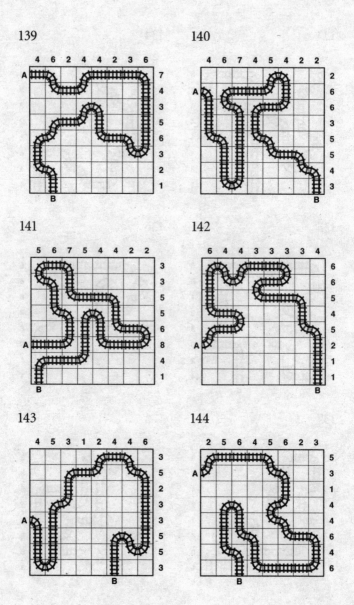

140

141

142

143

144

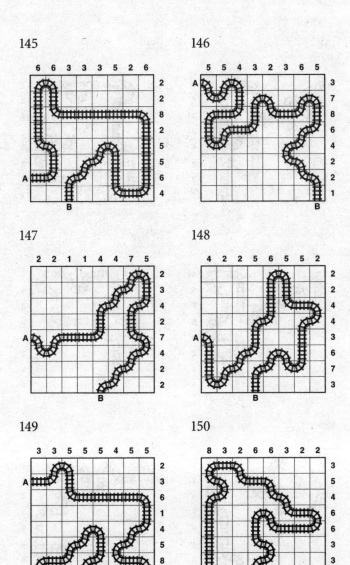

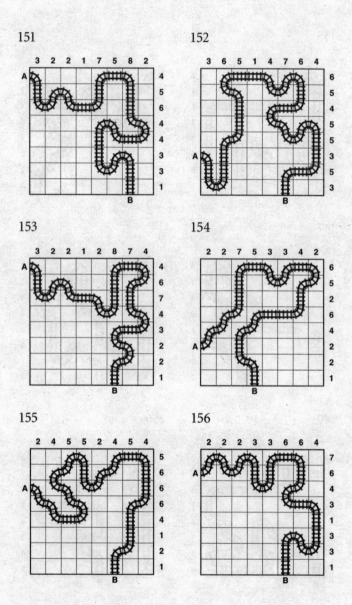

151

152

153

154

155

156

The Times Train Tracks

157

	6	4	8	3	5	3	3	2	
									2
									3
									5
									8
									6
									5
									4
									1

158

	3	4	5	4	2	2	4	2	
									4
									4
									2
									3
									5
									2
									2
									4

159

	1	3	8	1	5	3	6	2	
									5
									4
									2
									2
									4
									6
									2
									4

160

	2	1	3	7	7	5	3	4	
									5
									3
									2
									4
									5
									4
									5
									4

161

	2	1	3	5	3	7	3	8	8	7	
											3
											7
											3
											4
											6
											5
											5
											4
											3
											7

162

	3	4	2	2	4	5	8	5	3	6	
											10
											8
											4
											4
											2
											4
											5
											4
											1
											1

Solutions

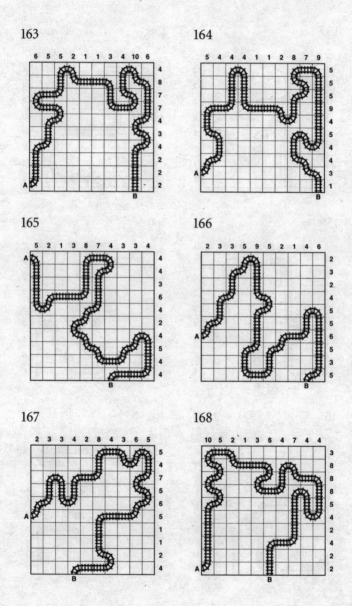

The Times Train Tracks

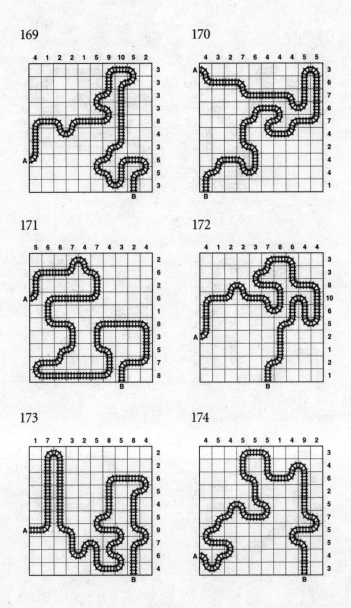

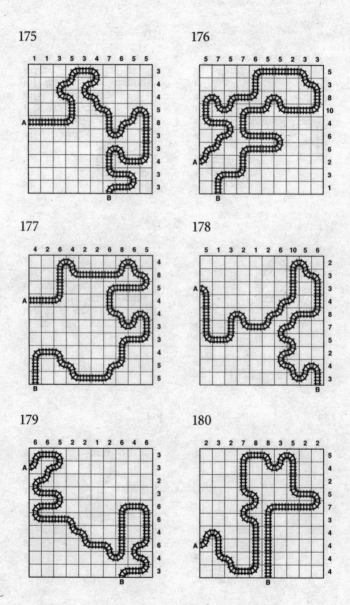

The Times Train Tracks

181

	2	1	2	3	5	7	7	9	2	4

A

5
7
3
3
3
1
6
5
6
3

B

182

	6	8	4	4	1	4	6	3	4	3

8
4
3
5
4
3
4
4
3
5

A

B

183

	5	7	4	3	2	1	6	5	4	5

3
3
5
7
7
5
6
2
3
1

A

B

184

	5	5	6	2	4	5	6	2	3	2

A

3
3
3
3
6
6
4
5
2
5

B

185

	5	5	3	3	4	6	3	9	4	2

4
4
6
10
4
7
2
3
3
1

A

B

186

	9	7	6	1	4	5	3	1	3	5

5
3
3
4
4
9
2
5
5
4

A

B

Solutions

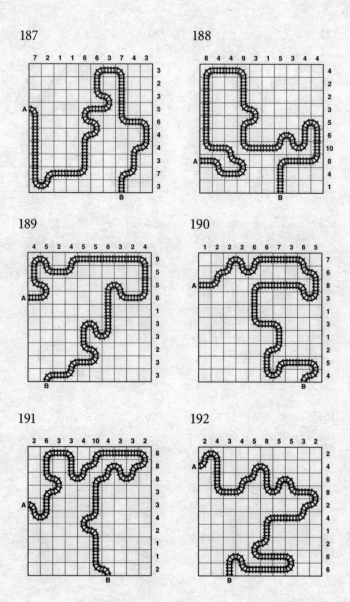

The Times Train Tracks

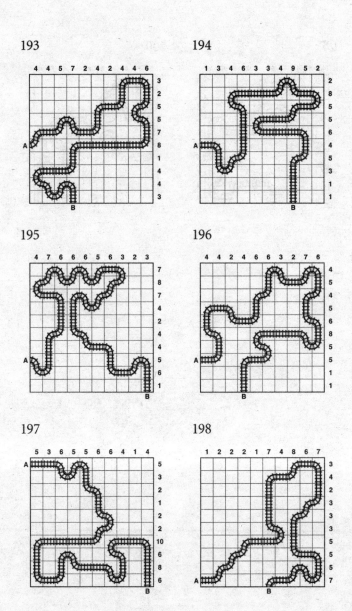

193

4	4	5	7	2	4	2	4	4	6

3
2
5
5
7
8
1
4
4
3

194

1	3	4	6	3	3	4	9	5	2

2
8
5
6
4
5
3
1
1

195

4	7	6	6	6	5	6	3	2	3

7
8
7
4
2
4
4
5
6
1

196

4	4	2	4	6	6	3	2	7	6

4
5
4
5
6
8
5
5
1
1

197

5	3	6	5	5	6	6	4	1	4

5
3
2
1
2
2
10
6
8
6

198

1	2	2	2	1	7	4	8	6	7

3
4
2
3
3
5
5
5
7

Solutions

199

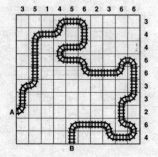

200

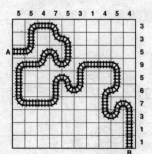